FORUM FOR SOCIAL STUDIES

FSS Studies on Poverty No. 5

Issues in Urban Poverty:
Two Selected Papers

Papers by
Daniel Kassahun
and
Meron Assefa

Addis Ababa
Forum for Social Studies
July 2005

ISBN: 1-904855-73-3
ISBN-13: 978-1-904855-73-6

Layout by: Mihret Demissew

Contents

Figures

Tables

Boxes

Introduction

In June of this year FSS hosted a panel on *Urban Poverty* at the Third International Conference on the Ethiopian Economy organized by the Ethiopian Economic Association (EEA) and held here in Addis Ababa. This was the third panel FSS has hosted at EEA's annual conferences in the last three years. The two papers published in this issue of our poverty studies series were originally presented at the panel. They have been revised and edited following the discussion at the panel and comments from the audience.

The urban sector has been largely neglected by researchers and policy analysts in this country, and debates on poverty often ignore the urban sector. While it is true that the overwhelming majority of the country's population is rural and the agricultural sector is an important contributor to the country's GDP, the heavy "rural bias" reflected in development policy, research and public debate will in due course come to distort our perception of the existing reality. If indeed the goal of development is to reverse the dominance of the rural and to place the urban sector at center stage, then it is important to begin to balance the focus of our attention. In some instances, and as Daniel's study published here suggests, it may be difficult to separate the rural problem from that of the urban.

FSS has in past been "guilty" of the rural bias noted above, but we are hoping to redress this by giving greater attention to the urban sector. Both papers are the work of FSS' research staff and they reflect work in progress rather than finished products. Daniel's paper is a study of the charcoal economy in the Meki-Zwai area, a subject that has not attracted sufficient attention. It examines the production, marketing and consumption of charcoal in both rural and urban areas and tries to show how this affects rural and urban poverty and the environment. Here there is a close link between the rural and the urban (the urban being medium sized towns in the area). Meron's study was undertaken in three *kebelles* in Addis Ababa and the focus of her analysis is the gender dimension of poverty.

Forum for Social Studies
July 2005

i

1

Some Aspects of the Charcoal Economy and
Its Impact on Poverty and the Environment

Daniel Kassahun

1. Introduction

In Ethiopia, urban areas contain 15 percent of the total population but are growing at a fast rate (> 4 percent). This is due to the combined effect of natural increase and increasing rural-urban migration. Such rapid growth of urbanization has induced an ever growing necessity for household energy.

The African 10 Year Framework Program on Sustainable Consumption and Production, which has been approved by the African Ministerial Conference on Environment, recognized urban centers as the strategic entry point for the promotion of sustainable consumption and production. In the conference, energy was one of the four principal thematic areas which constituted the African 10 Year Framework Program. This is because energy is related to the most pressing social, environmental, economic and security issues which affect sustainable development.

In Ethiopia, biomass is the main source of energy, which accounts for over 90 percent of the total energy consumption (Yacob, 2003) which, at 0.02 tones of oil equivalent (toe) per capita (Bereket *et al.*, 2002), is among the lowest in the world. Paradoxically, the country is endowed with tremendous potential of renewable energy resource (Table 1). Expenditure on firewood accounts for 38 percent of the households' energy budget, followed by electricity (20 percent) and kerosene (19 percent). While charcoal accounts for 17 percent, dung cakes account for 5 percent of the household energy budget (Bereket et al., 2002). This is an overall data where there is marked variation at different places.

Table 1.1 Energy sources of Ethiopia

Source	Exploitable reserve	Units	Exploited percent
Hydro power	30,000	KW	3.3
Solar insolation/day	5.3	KWh/m^2	~0
Wind speed	3.5-5.5	m/s	~0
Geothermal	700	MW	1.2
Natural gas	76.5	Billion m^3	0
Coal	13.7	Million tons	0

Source: Adopted from Wolde-Ghiorgis (2001)

Heavy dependence on biofuels has given rise to concerns that the forest resource could be decimated, and this calls for major interventions. Rapid population growth in the absence of energy substitutes for traditional energy sources are accelerators of forest depletion (Wolde-Ghiorgis, 2001). This brings about the physical and economic deterioration of biomass resources and hence higher household expenditures of labor and time. As the biomass stocks are diminishing fast, Yacob (2003) has argued that Ethiopia should embark on a radically different energy path.

Anti poverty initiatives in Ethiopia have targeted rural areas as government is overwhelmingly pro-rural. Efforts to address the unique problem of urban poverty have been negligible. Within the urban milieu, small towns, which are catalysts of rural development, have not received the attention they deserve. This is testified by the fact that most of the development research in Ethiopia has focused on large and primate cities (e.g., Ciamantini and Patassini, 1996; Jogan and Patassini, 1996; Dierig, 1999; Bereket *et al.*, 2004; Girma, 2004). On the other hand, within the rural areas, energy development has not received a fair share of public investment in comparison to education, rural road construction and health.

Household energy consumption is known to be intimately related to income, resource availability, and environment. People living in poverty are often disproportionately victims of environmental effects related to energy. Hence, the prevailing patterns of energy consumption among poor people tend to worsen their misery (Dasgupta, 1993). This is partly because in the traditional wood-burning stoves, only 10 to 15 percent of the released energy is useful. By contrast, 50 to 65 percent of the energy from liquid petroleum gas (LPG) is useful. In Philippines, the urban poor pay US$1.79 per kilogram of oil equivalent (kgoe) for their cooking needs, whereas the rich pay only US$0.66 per kgoe, mainly because the poor uses firewood and the rich LPG (Barnes, 1995). Such facts are further exacerbated by the fact that poorer households have larger family size (MoFED, 2002).

Energy use by households varies across quality and efficiency called "energy ladder" (Figure 1). Increasing cleanliness, efficiency, cost, and convenience are shown in the direction of "increasing prosperity". The opposite is "declining prosperity", which could result from depletion of existing energy sources or due to crippled economic status of households. While firewood and cattle dung belongs to the lowest rung, charcoal, coal, and kerosene represent the next rungs up the ladder of energy. Electricity and LPG are at the highest rungs. In Ethiopia, electrification is woefully inadequate, and only 12 percent of urban households are electrified. This shows that the majority of Ethiopians use inefficient and relatively more polluting energy sources and most often are forced to engage in ecologically damaging activities. Hence, any anticipated economic development opportunity would be slowly absorbed by the poor. This is because, any additional income in the household would first go to buffer the energy triggered deficit of household economy.

In many parts of Ethiopia, cattle dung is used as fuel rather than as fertilizer because of shortage of firewood (Senait, 1999; Hirut, 2000; Woldeamlak, 2005). Patterns of changing energy sources, which is from the use of fuelwood to cattle dung, had resulted in various repercussions in Ethiopian highlands. Hirut (2000) noted a reduction in energy use, changing of plates for baking, changes of the type of food prepared, and reduction of the frequency of cooking.

Subsidies for commercial fuels had been widely recommended by scholars, which are believed to extend their use by poor households and hence lower the proportion of income spent on energy. However, it is the middle-class and better-off households who reported to garner a disproportionate share of the benefits because they can afford to use much more energy than poorer households (Douglas, 1995).

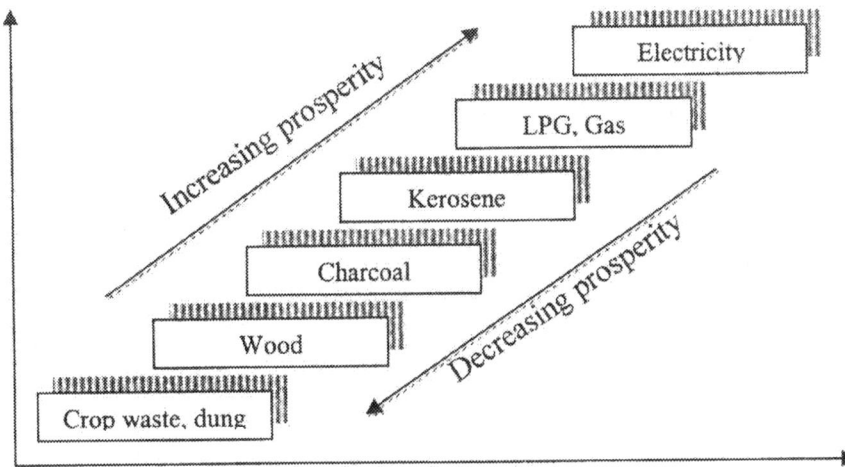

Figure 1.1 The household energy ladder. Modified from Meikle and Bannister (2003)

Energy extraction and its pattern of use has strong environmental consequences at local, regional and global levels. While energy interventions are believed to help poverty alleviation and environmental protection, very little research has been undertaken in this regard (Meikle and Bannister, 2003). As the linkages between poverty and environment are complex, they require locally-specific studies to fully understand them.

Many scholars agree that poverty and environmental management are intimately intertwined. People living in poverty are forced to overuse environmental resources for their daily survival and are further impoverished by the resulting degradation (Figure 2). The unsustainable nexus between population and environment therefore leads through a downward spiral into a poverty trap. The perception of peasants about environmental dynamics shapes the way people manage the environment.

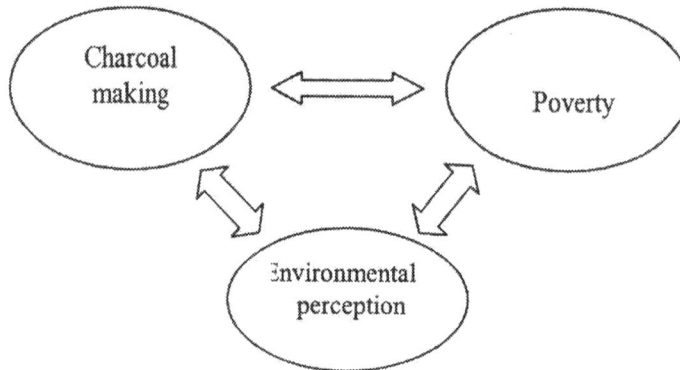

Figure 1.2 Conceptual interrelationship between charcoal making, poverty and environmental perception

As shown in Figure 2, degradation of the environment due to exploitation of biofuels and its impact on food productivity could make households environmentally conscious. However, as most farmers are living at the brink of starvation they might be forced to exploit the remaining environmental resource while knowing its damaging effect. Currently, there is a growing hitch of 'energy-crisis' in most parts of Ethiopia. For instance, in Makale and Adigrat towns, people are forced to use roots of trees for fuel. In the latter, the price of 50 kg of firewood is almost equivalent to the price of 50 kg of grain (EPA, 1998).

The energy consumption pattern is dependent on demand and supply. So far, studies in Ethiopia have focused either entirely on the demand side (Bereket *et al.*, 2002; 2004) or in the supply side (Senait, 1999; Hirut, 2000; Woldeamlak, 2005). Studies integrating the supply and demand sides are overlooked. While population, along with generated energy demand, is growing fast, the supply aspect of charcoal (in the form of volume and accessibility) is diminishing alarmingly. This brings the necessity of integrating the supply and demand side.

Charcoal has a higher demand directly from small towns and indirectly from far away big cities. Urban centers act like magnets which disproportionately drain the non renewable natural resources from their adjoining and distant rural environs. Hence, the rural ecosystem is overburdened not only from the growing pressure of population of the locality, but also from far away cities like Addis Ababa. In response, urban centers offer a range of economic possibilities and niches for surrounding rural villages. Compared to other lowlands, the central part of the rift valley is crisscrossed by a transportation network, which makes the wood resource to be sapped at an accelerated rate. The fact that at least one sack of charcoal is legally allowed for any vehicle passing through the region has created an insatiable demand for charcoal. This is due to the ever increasing number of trucks passing through the region. However, the role of small towns as a node and conduit for charcoal consumption is overlooked by researchers.

As shown in Table 2, it is not the rural areas which consume charcoal. Rather, the better-off households residing in Addis Ababa which might have triggered the influx of charcoal to the cities.

Table 1.2 Types of domestic energy used by households (% of households)

Source	Country level	Urban	Rural
Collected firewood	67.78	16.84	76.00
Purchased firewood	7.69	41.22	2.28
Charcoal	1.19	7.97	-
Kerosene	3.28	21.78	0.30
Butane gas	0.22	1.25	-
Electric	0.38	2.19	-
Dung cake/leaves	15.99	6.38	17.54

Source: Adapted from MoFED (2002)

It goes without saying that the current pattern of household energy use in Ethiopia varies across space and time. One can easily compare the problem in a highland-lowland dichotomy. While most parts of the highland areas were subjected to a long history of settlement and agriculture, most of the natural resources have undergone severe degradation (e.g., Hurni, 1988; Mesfin, 1991; and Pankhurst, 1992). The natural vegetation in most of these areas had vanished a long time ago except in inaccessible areas (like steep valleys) and distant areas from transportation routes.

Presently, the wood and charcoal markets are located in the lowland-highland interface. In most parts of the highlands, the energy ladder is stuck at the lower energy rungs, where the use of cattle dung and crop residue is most common. Woldeamlak (2005) has reported 100 percent energy consumption in the from of cattle dung in northwestern highlands. In contrast, in the lowland parts of Ethiopia, the man-land ratio has been extremely little, and hence the intensity of biofuel exploitation has been modest. However, these areas are ecologically fragile. The soils are highly erodible. Once the trees are removed, the soil would be exposed to erosive rainfall and blowing winds. The dominance of sandy soil in these areas is reducing the water-holding capacity of the land and therefore making it susceptible to crop failure. These characteristics make the area under consideration remain with little carrying capacity, gave it a higher risk of food insufficiency, and promote recurrent drought and famine.

In brief, the existing studies of domestic energy related issues in Ethiopia could be characterized either as lacking demand-supply integrity, or failing to examine the energy situation in small towns. They disregard the seasonality of charcoal making and are incapable of capturing the relationship between energy making, poverty and environmental awareness.

In view of these problems, this study specifically aims:
1. To assess the supply aspect of charcoal in the ecologically-sensitive rural environment;
2. To examine the demand and consumption aspects of charcoal in small urban context;
3. To identify the nexus between household energy, poverty and environment; and

4. To generate debatable issues on household energy and contribute to better policies, identification of priority areas of research and understanding of the house energy problem.

The study area lies roughly between 8^00 to $8^030'$N and $38^030'$ to $39^015'$E. Administratively, it is located in *Dugda Bora wereda*, East Showa Zone, Oromia Regional State. *Dugda Bora* is one of the 12 *weredas* found in the zone. The area is situated between 100-130 km south of Addis Ababa. While the total population of the *wereda* is estimated 190 375, the total area is 1,459.5 km^2, which makes a density of 130.4 person/km^2. Lake *Ziway* and *Koka* are found in the *wereda*.

The lakes and rift valley portions of the semi-arid zones have a crop growing period of between 46-60 days. Natural tree species found in this part include *Boswellia papyrifera, Acacia segal, Acacia senegal, Acacianilotica, Zyzyphus* spp, *Diospyros mesqiliforms, Exytenanthera abyssinica, Balantines aegyptica* (EPA, 1998). Precipitation is low, 739 mm per annum, and erratic in most of the rain-fed areas. There is a high coefficient of variation with regard to amount, onset and cessation of rainfall.

Topographically, the *wereda* is dominantly of plain surface, where 82 percent of the land has a slope of 0 to 10^0 and 3.4 percent lies in a slope range of 10 to 34^0. The general elevation ranges from 1600 to 2000 m above sea level.

The study area was selected as a case study due to the following reasons:
(a) It represents a rapidly degrading area overlaid with widespread poverty;
(b) The area is famous for charcoal production in the country;
(c) The accessibility of the district to conduct fieldwork; and
(d) The presence of ongoing activities addressing urban poverty and land degradation which could shed some light on the link between the degrading environment, poverty and local environmental perception.

2. Research Methodology

Two different but complementary approaches were applied: viz., analysis of the demand and supply aspects of charcoal. Correspondingly, two groups of respondents were chosen: rural households (*biofuel makers*) and urban households (*biofuel consumers and distributors*). The field survey was conducted in March 2005. The rural villages (kebeles) selected for study were *Tuchi Sumeya* and *Dalota Mati* and the number of households in them was 392 and 354, respectively. They are found in the neighborhood of *Meki* and *Alem Tena* towns, 7 and 20 kms to the east of the towns, respectively. Both kebeles are inhabited by Oromo people, who are mainly Orthodox Christians. Here, Focus Group Discussion (FGD) was held with various age groups to assess the patterns of charcoal production, poverty and their perception on local environment.

Plate 1. Focal Group Discussion with a peasants composed of elders, adults and women

A structured household survey was carried out in the two urban centers of the *wereda*: *Meki* (in 3 *kebeles*) and *Alem Tena* (in 2 *kebeles*). The two towns are among the 32 towns found in East Oromia zone. Systematic sampling technique was employed and a total of 100 urban households were involved where each *kebele* was represented by 20 households. In each *kebele*, every 20th house number was chosen and but where the chosen households were found reluctant or not available, the immediate neighbor was selected.

The survey questionnaire was administered by five field data collectors, one in each *kebele*, and one supervisor for each town. Most of the assistants had previous experience in similar household surveys. Before the survey, a half day training was conducted. SPSS, a statistical software, was used to analyze the questionnaire-generated quantitative data. Descriptive statistics and ANOVA were used to summarize and assess the interrelationship between selected socio-economic variables. In addition, persons who have immediate attachment to charcoal were employed for in-depth informal discussion, which include leaders of *kebeles*, police officers, and mayors of the two towns.

3. Charcoal Production and Its Impact on Poverty and the Environment

The ever-growing human- and natural-induced stress on the rift valley region is disrupting its delicate ecological equilibrium. The fast rate of environmental degradation is partly due to the higher susceptibility of soils for wind and water erosion. Despite the predominance of plain topographic surfaces in the locality (which might have deterred the accelerated runoff), the depletion of their biotic cover had enabled the proliferation of gullies. The once densely populated *Acacia sp* trees have dropped to a present density of about 10-15 trees per ha.

Respondents of both rural *kebeles* recount that about four decades ago, there had been dense vegetation in the region. To them, the deforestation process was begun in the form of selective clearing during the reign of Emperor Haile Sillasie. Initially, big acacia trees were specifically targeted. Before that, even lost animals were not easily found in the woods. People residing in the locality were very few and engaged dominantly in pastoralism.

It was around 1964-65 that settlers called *bale wuleta* (*fevered citizens*), who were said to have come from the 'north', introduced charcoal production, popularized arable farming, and enabled widespread settlement in the present locality. These *bale wuletas* were also known as "charcoal producers". This historical account is also shared by the towns' *kebele* officials. While the *kebele* officials explain that *bale wuletas* were retired military members, the rural peasants believe they had come from *Sudan*. The word *Sudanie* was a derogatory term used at that time to refer to people who were engaged in charcoal making. Rural respondents explained that the then 'charcoal producers' were operating along the entire rift valley extending to the Ethio-Kenyan border.

Plate 2. Acacia is the most preferred tree for charcoal making. Presently they are barely found in farmstead or homesteads and are cut during moments of food scarcity.

Due to the ever-growing demand by settlers and the corresponding worsening of the environment, the community got involved in charcoal making as a fast income generation scheme. Respondents expressed that from a single acacia tree about 2-3 sacks of charcoal could be harvested. Such activity unavoidably inflicts collateral damage, where many non charcoalable trees would be lost for making pavement, firing trunks, covering, etc. These days, due to the depletion of acacia trees, rural people have started to use every type of natural vegetation, which used to be inferior for charcoal making. Even tree stumps are used for charcoal making. When the settlers colonized the present rural areas, charcoal production was undertaken only during special moments of hardship. Recently, however, it has become a common activity, not only for cash for purchasing of grains, but also for participation in traditional socio- economic institutions like '*ekub*' and '*edir*'.

Peasants admit that making charcoal enables them to earn lucrative income, which tempts farmers to cut down more trees. The immediate conclusion one might draw from this is that more and more trees are being cut down. The household energy source in the rural locality used to be charcoal and firewood but it is very recently its quality has deteriorated and people now use cattle dung. Peasants cut acacia while knowing the tree does not re-vegetate after felling. On the other hand, the introduction of donkey-pulled carts in the locality has increased the production and transportation of charcoal, wood, tree stumps, and other commodities to the main urban centers. This has accentuated not only the deprivation of soil cover but also the unsustainable mining of soil nutrients.

In many cases, the demand for woodfuel is believed to be a motivation to plant trees to develop agroforestry and community forestry. However, the eucalyptus tree, which matures fast and serves various purposes in many highland parts, is not accustomed in the locality. Hence energy-crisis is a budding phenomenon in the region.

While peasants know the effect of rapid population growth on the environment, the uncontrolled birth rate is considered 'natural'. Gullies are rapidly expanding and eating their farmlands. Peasants themselves witnessed that the land is very susceptible to sheet and gully erosion. The prime factor for the depletion of natural vegetation in the locality, according to respondents, was clearing of vegetation for cultivation purposes, which was practiced by all settlers, followed by charcoal production, which was undertaken by poor peasants only. This belief goes in line with the assertion of Eckholm (1984) that in most countries, forests are disappearing not because people want to burn the trees, but because they want the land under the trees for agriculture.

In charcoal making, younger farmers used to dominate the activity. Subsequently, all segments of the rural population including females are actively involved in the activity. Wives usually "steel" small amounts of charcoal from what has been produced by their husbands and sell it in towns. Discussants noted that poverty of the local people, which is partly induced by environmental degradation, has further instigated farmers to degrade the remaining woody resource.

Plate 3. Badlands: While the topography is dominantly of gentle slope, soils of the study area are very sensitive to erosion. Once the natural cover is removed, erosion is rapid. An attempt is made to plant trees, but their intended role looks very bleak.

As a consequence of natural degradation, respondents stated the local weather has shown evident change through time. The number of animals owned by a household has been declining while the overall number of livestock has increased due to the rapidly growing number of households. This has brought scarcity of animal feed. In earlier times, the locality was famous for keeping large number of livestock. Possession of 50 animals was considered "very little". Nowadays, the carrying capacity of the area is reported to have declined. Respondents explain that half a hectare of grazing land does not suffice for a pair of cattle to feed on. Farmers explain that it is because the soil has deteriorated and even the quality of feed type has shown qualitative change.

Plate 4. Remnant erosion features: the picture witnesses the magnitude of soils lost in the area

It is interesting to note that charcoal making is mostly undertaken during summer (Figure 3). Discussants correspond this period with shortage of seed, food, medicine, and entanglement in "shark" loans. *Kebele* officials in the two towns also confirmed the seasonality of charcoal movement. Since farmers are living on the edge of survival, they are liable to have short time preference and high discount rate. These farmers rationally choose to depreciate their soil/tree resources when survival is at stake especially during summer. Such seasonality of household food scarcity was also noted by Mesfin (1997), where hunger is not a particular episode like flood, earthquake, or fire incidence. He explained that in order for eventual hunger to occur, the process builds up over several months. Starvation commences as of February immediately after the harvest period. In early summer, it reaches climax even in good years. This period is commonly known as "pre cropping starvation".

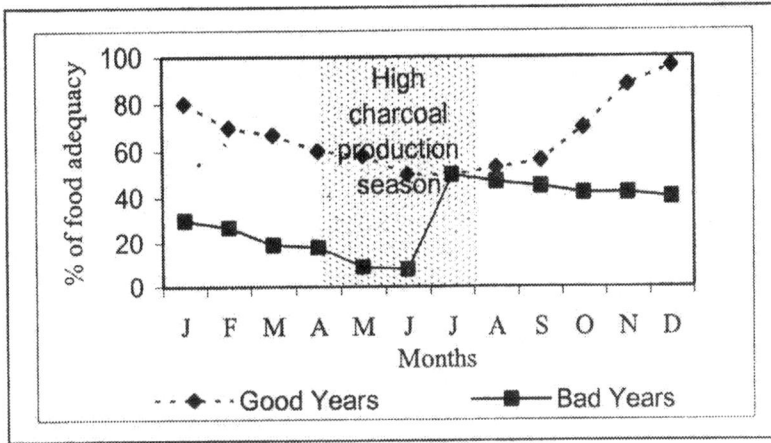

Figure 1.3 Seasonality of charcoal production and food adequacy in the rural household (Source: synthesis of Mesfin (1997) and data from the study)

Plate 5. Lake Koka: due to worsening environmental factors, the lakes in the study area are retreating fast, and people are colonizing such receding lands.

In figure 4, controlling variables are the intensifying poverty through decades and seasonality of food security, while the state variable is the intensity of charcoal making. Despite the general decline of charcoal making through years, there are special episodes of peak and dip depending on the long and short cycle drought incidence and the intermediate good years.

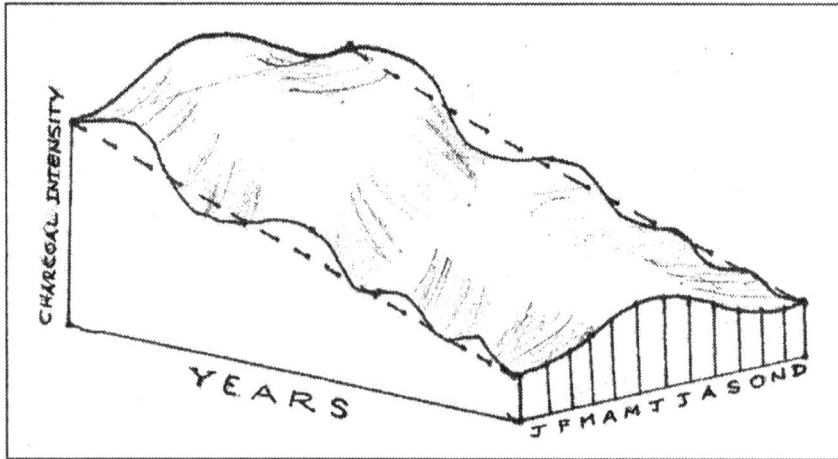

Figure 1.4 A hypothetical deductive modeling of feedback between charcoal making, growing intensity of poverty and seasonality

Respondents believe that the locality has sustained an intensified environmental degradation through time. This process has sharpened the social conflict with neighboring localities, which dates back to the time of Emperor Haile Sillasie. Nowadays, the conflict has reached a climax and residents fear that the worst may happen. It is a shared idea among respondents that efforts to minimize forest degradation should not be localized to a specific area; they liken this to partial ornament (*gintil get*). It should rather encompass large geographical areas, where the environment could be managed holistically. In response, there had recently been several attempts to make negotiation with neighboring localities (*Ziway-Bora Dugda*), but it did not bear the anticipated fruit. Respondents in *Tuchi Sumeya* rural *kebele* said "the conflict declined once the forests in the buffer zone got completely depleted in 1995".

Town officials believe that charcoal production in the rural area is having a series impact on the environment of the towns, too. These days, the temperature is hotter, the rain is less, and the susceptibility of the area to drought and desertification is higher. They explain that the focus of the poor hinges more on daily livelihood than on long term benefits. As explained by town officials, due to the magnitude of poverty, the rural poor say "let me eat today, and die tomorrow".

Rural respondents pointed out that the demand for charcoal has shown a marked increase through time, even from far away towns. These days, the demand for charcoal is met mainly by distant peasants in Arsi zone. On big market days more than a hundred donkeys laden with charcoal and firewood flock to towns from Arsi. An individual could bring 5-6 bundles to market by donkey. This is only what is sold in open markets. There is a substantial amount which is transported on the main highway without passing through the towns.

For a long time, the charcoal market used to operate openly. Nowadays, it is undertaken secretly, where the brokers make deals by traveling house to house in rural *kebeles*.

According to one official[1], petty traders and restaurant owners usually go to rural *kebeles* to convince farmers to cut homestead or farmstead trees for sale. These business men have a network stretching up to Addis Ababa.

Compared to the adjacent eastern and western plateaus, natural degradation in the rift valley is a very 'recent' experience. However, the area has recently undergone degradation at unprecedented rate. Rapid population growth, fast land cover/use change, swift depletion of woodlands, alarming shrinkage of lakes, etc took place especially in the last 3 to 4 decades. Hence, the vulnerability of the locality to meteorological anomalies has been aggravated through time. When this factor is interwoven with the growing scale of poverty, the problem gets complicated.

Rural households believe that soil degradation, which is the result of land cover depletion, is responsible not only for blocking many of water points (*sink holes*) in the locality but also for the sedimentation of Lake Ziway. Recently, the lake has considerably shrunk in volume. Some tree species, which were adapted to grow under the lake ecosystem, have now been exposed from the water. These buoyant and squashy trunks of trees, locally called *boffa*, are now widely exploited for craft making like stools.

In relation to environmental impact, rural respondents recount that in the old days malaria caused very mild sickness. The local adage used to be "*Ibiddi garaa beekan kobota; Dhibeen garaa beekan busaa dha*", which literally means "as *there is no fire sparkled from a dung, there are no sickness emanated from malaria*". These days, yellow fever is a major killer.

Due to the worsening of the environmental problem and the lesson obtained by peasants after experiencing the challenge, the following changes were noted in the rural localities:

- peasants have stopped harvesting trees together with its root part;
- peasants have started using earthen bricks, instead of wood, for the construction of houses and barns; and
- for making homestead fences, initially freshly pruned trees are used, and upon decaying, they will be used as firewood.

Due to unabated environmental deterioration in the locality, the following impacts were outlined by the discussants:

- The rain has decreased in amount, irregular in its periodicity, falling non-contagiously;
- The intensity of the wind blown erosion has increased;
- The risk of vegetation loss has increased;
- Local temperature levels have risen;
- Desertification symptoms have grown strong;
- The hazard and intensity of soil loss has increased;

---

[1] Ato Shumi, head of natural resource, *Dugda Bora wereda*

- Sedimentation and water pollution levels of the lake have increased;
- Livestock have suffered from recurrent drought. For instance, in 2003 there was severe drought and many livestock died. The milk productivity of the cattle has significantly declined;
- The period of land fallowing has significantly shortened and most farmers have abandoned it;
- Social conflict with the neighboring localities has intensified;
- Drought used to occur at very long intervals, then increased to every 4-5 years, and very recently, it is occurring every other year. As noted by the *wereda* natural resource expert, massive degradation of forest took place in drought periods as to generate income through selling charcoal and firewood;
- Malaria incidence has intensified; and
- Peasants have become pessimistic about the future environmental wellbeing and some are afraid that the land may cease to grow crops completely after 2 to 3 years.

Many of the respondents have witnessed discernible ecological change during their life time, and they label the trend as 'alarming'. They are afraid of the likelihood of desertification in the near future which could be devastating.

4. Charcoal Consumption and Its Impact on Poverty and the Environment

The population of *Meki* and *Alem Tena* towns is 34,914 and 12,906, respectively. Of a total of 100 sample households included in the survey, females constitute 61 percent. Nearly half of the sample households were born locally and are very familiar with the pattern of energy use in relation to environmental change in the locality. While 68 percent of sample households possess own houses, 21 percent are tenants of the *kebele*. Only 10 percent of the households have rented houses from private owners.

Table 1.3 Summary of household parameters in *Meki* and *Alem Tena* towns

	N	Min	Max	Mean	Std. Dev
Age of respondents	100	18	80	41.62	14.42
No. of rooms in households	100	1	15	3.84	2.91
No. of lights in households	100	1	15	3.69	3.11
Weekly consumption of charcoal	89	0.50	30.00	8.47	5.79
Av. monthly payment of electric	99	2.00	150.00	23.06	22.38

Source: Survey data, 2005

Meki and *Alem Tena* towns do not fulfill the criteria of urbanity the strict sense. Substantial proportions of residents are engaged in agriculture in the nearby rural areas. The majority of dwellers are engaged in petty business and as daily laborers. According to the mayors and *kebele* officials of the two towns, people engaged in cultivation are working either on their own land or land contracted from neighboring rural areas. Compared to *Meki*, most urban farmers in *Alem Tena* produce their annual food by themselves. This is because contracting rural land is considered "very cheap". One of the reasons for involvement of urbanites in agriculture is that many peasants who used to live in the rural area built residential units in towns so as to let their children attend formal schools. Besides, possession of houses in towns is considered as a sign of prestige in the rural community.

Officials noted that the towns' population is rapidly increasing, mainly because of immigration from other areas and high local fertility. The influx of immigration was intensified especially after the inception of vegetable farming in the locality. However, the majority of the people in the town live in abject poverty. As the mayor of *Alem Tena* town put it, *"the town is the seat of the poor"*. Most inhabitants are engaged in middle and lower economic activity, on a daily basis. Besides, there are a substantial number of unemployed citizens in the towns.

Plate 6. Charcoal ready to be dispatched to major cities like Addis Ababa: each vehicle passing through the region is allowed to buy and carry one sack.

The charcoal, which comes to the towns, is not wholly consumed by local people. Rather, the bulk of it is collected by wholesalers for transportation to Addis Ababa by big trucks. These urban areas serve mainly as collection and transit nodes for shipment to far destinations. While the majority (> 65 percent) of Addis Ababa residents uses kerosene for cooking, almost all charcoal from the study area (also from other charcoal producing areas) is destined for Addis Ababa.

All respondents of the two towns employ electric mainly for home light. While 70 percent of households are EELPA's customers, the remaining 30 percent acquire electric power indirectly via regular customers. In some instances, the length of wires extended to those houses could stretch up to 100 meters from the source houses, and the local rate for such extension line is fixed at 10 birr per bulb. Although the variable rate is intended to benefit low electricity consumers (Figure 5) and thus the poor, this strategy fails to take into account the energy management strategy of the poor, many of whom share the cost of electricity connection in order to save money. This means that the combined electricity consumption quickly reaches the higher unit cost rate.

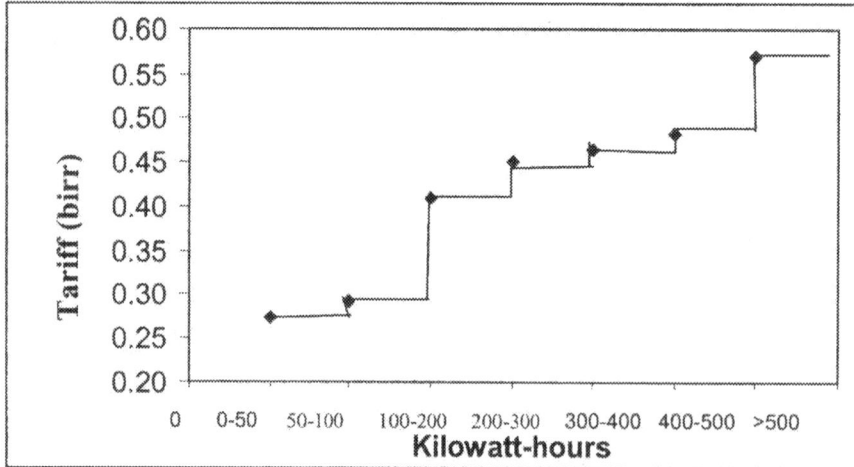

Figure 1.5 Stepped electricity charges (2005)[2].

Due to the exorbitant cost of electricity and the unaffordability of some electric devises, only 25 and 4 percent of households use electricity for television and refrigerator, respectively. None of them can afford electric supply for ironing, stove and *mitad*. While 96 percent of respondents bake *enjera* by firewood, 94 percent of respondents use charcoal as a primary source for cooking food (other than *enjera*) followed by firewood (35 percent), kerosene (25 percent), and dung (18 percent). The town officials also confirm that there is severe shortage of household energy, especially firewood and the majority of people currently use cow dung and kerosene for cooking. This trend is substantiated by Bereket *et al.* (2002), where the national average cost of electricity is more than 5 times the cost of using kerosene for stove and lamp.

Despite the high (65 percent) involvement of urban poor in petty charcoal retailing, their charcoal consumption at home household is very negligible. Less than 10 percent consume charcoal for their cooking, while the majority (> 70 percent) uses cattle dung. Besides, the ANOVA test demonstrates that the number of lamps in households is positively and significantly ($p < 0.01$) related to charcoal consumption.

Appreciation of the constraints and of its environmental impact on the supply side of the energy source, especially the biofuels, could influence the pattern of its consumption in the towns. Almost all members of urban society employ charcoal and the majority (78 percent) buys it from small local markets (*gulit*). Table 4 shows that nearly half of the respondents perceived that charcoal, firewood and cattle dung obtainable at urban markets are produced from their immediate rural neighborhood. The remaining half believes that the biofuels come either from distant localities or have no knowledge of its place of origin.

---

[2] *Data directly obtained from EELPA*

16

Table 1.4 Perception of respondents on the sources of biofuels available in town markets (% of respondents)

Source origin from:	Immediate surrounding	Far away locality	I don't know
Charcoal	55	36	7
Dung	22	53	14
Firewood	52	44	3

Source: Survey data, 2005

Respondents believe that, from time to time, the volume of charcoal arriving at the markets decreases (77 percent); with a corresponding price increase (93 percent); and the number of individuals engaged in charcoal selling have decreased (70 percent) (Table 5). This shows that the public is generally aware of the ongoing deterioration of the surrounding environment, and its impact has on the price, job creation, and affordability of charcoal. With respect to the rate of deforestation, 76 percent of respondents put it as "very high", and 17 percent as "high". More than 80 percent of residents have lived in the locality for more than 20 years.

Table 1.5 Perception of respondents on temporal aspects of charcoal (% of respondents)

Time to time pattern in the:	Increased	Decreased	Constant
volume of charcoal	15	77	5
price of charcoal	93	2	4
no. of charcoal sellers	27	70	2

Source: Survey data, 2005

As discussed in Section 3, on the supply aspect of charcoal in rural villages, the seasonal nature of charcoal inflow is substantially confirmed by households in the two towns. This result goes in line not only with their rural counterpart but also with town officials. About 75 percent of respondents confirmed that there is voluminous influx of charcoal to the towns during summer season. *Kebele* officials and police officers explain that, during the summer season, the majority of peasants exhaust their food and cash resources.

The survey further showed that the level of household income has decisive role on the choice of energy type (Table 6). Had the household income been substantially increased, the majority of households (91 percent) might have shifted to electric use for cooking leading to a substantial decline (76 percent) in cattle dung consumption. A moderate decline (61 percent) of charcoal could be attained in such circumstances. Strangely enough, the consumption of fuelwood seems to be insensitive to increasing income levels. The result implies that improving the economic status of the urban poor would ease the unbalanced burden on the biofuels and thereby reduce the rate of natural vegetation depletion and soil loss. This finding goes in line with results obtained by Bereket *et al.* (2004) that dependence of households on biomass significantly decreases with increasing income on aggregate level.

Table 1.6 Probable impact of income increment on biofuel consumption (% of respondents)

If income increases:	Increases	Decreases	Remain constant
Charcoal use	25	61	9
Dung use	12	76	12
Electric use	91	2	6
Fire wood	41	51	8

Source: Survey data, 2005

As noted above, urban households are very well aware of the source of biofuel and its impact on environmental wellbeing. Respondents have attributed the impact of charcoal production to various aspects of environmental parameters (Table 7). Most of them have attributed it to ominous desertification processes. This perception is further strengthened by rising temperatures and declining rainfall. Compared to their rural counterpart, urban respondents are less pessimistic on the impact of charcoals on soil loss (58 percent), wind erosion (66 percent) and loss of wildlife (54 percent).

Table 1.7 Perceived impacts of charcoal production on environment (% of respondents)

Charcoal making induces impact on:	Yes	No
Desertification	96	4
Soil loss	58	42
Loss on wildlife	54	46
Wind erosion	66	34
Temperature increase	87	13
Rainfall decrease	80	20

Source: Survey data, 2005

The environment of the locality has severely been affected since the 1984 famine. Officials of *kebele* 02 in *Alem Tena* justified that, "people are knowledgeable of the impact of charcoal on environment. It is their poor economic status that drives them to pursue a detrimental activity". Urban respondents perceive that the environmental degradation of the locality has a series impact on the lake ecosystem. The volume of Ziway Lake, the shallowest in rift valley, is shrinking, where cultivation is taking place in the receding part of the lake.

While acknowledging the impact of charcoal and other biofuel exploitation from neighboring localities, respondents had divergent insights on measures to avert the problem. While there are substantial proportions (61 percent) of households that accept a ban on charcoal production, 22 percent do not support the idea (Table 8). Nearly half the respondents believe charcoal production is controllable by the government, though they are afraid it may not be able to effectively accomplish it.

18

Table 1.8 Opinion of urban households on charcoal control (% of respondents)

Type of opinion:	Yes	No	I Don't Know
Should charcoal production allowed to continue?	22	61	15
Is government effective in charcoal production?	49	41	6
Is charcoal production in the locality controllable?	38	48	11
If electricity is subsidized, would charcoal demand decline?	75	14	9

Source: Survey data, 2005

Charcoal is useful source of income to most of the people in the towns. If its circulation is banned entirely, the mayor of *Alem Tena* believes that "many people, especially the poor, would get hurt". *Kebele* and town officials believe that poverty is strongly entwined with the charcoal business. "Unless a person has no source of income, why does he filthy himself with charcoal and ash"? It is the poorest farmers, mostly lacking oxen, who are mostly engaged in charcoal to sustain their life.

Police officials also believe that charcoal is a source of livelihood for many people in the towns. While wood (lumber) comes from far away *weredas*, charcoal is mainly produced in the nearby locality. Police also believe that most individuals who are involved in the charcoal production and retail are poor, who have no alternative occupation. As they pointed out, businessmen from other towns are not only collecting charcoal from petty vendors, but also go to rural villages, and persuade poor farmers to cut down the remnant acacia trees available in farmsteads.

Plate 7. Confiscated wood and charcoal at the *Wereda* Agriculture Bureau. Charcoal and wood production/distribution is "illegal". However, the control system is sporadic and the focus is after the trees are being cut.

Respondents had a varying opinion on the impact of banning charcoal production on different sectors of the society (Table 9). The majority (75 percent) have the belief that the poor would be the primary victims, whereas its impact on the rich is negligible (2 percent).

Table 1.9 Perceptions on the effects of banning charcoal production (% of respondents)

Total ban of charcoal affects:	Yes	No
Charcoal producers	58	42
Urban poor	75	25
Urban retailers	35	65
People in big cities	8	92
Business people	4	96
Urban riches	2	98

Source: Survey data, 2005

While acknowledging the ongoing resource degradation, partly induced by massive depletion of woodlands, respondents do vary in their opinion of the future wellbeing of the environment (Table 10). About 45 percent of respondents believe that the environment would rehabilitate in the coming years. This opinion contradicts with their rural counterparts, who are rather pessimistic due to the impact of changing environmental parameters on agricultural activities.

Table 1.10 Opinion of households on the future environmental wellbeing of the locality (% of respondents)

The future of environmental wellbeing:	%
Would rehabilite	45
Remain same	8
Worsen	39
Difficult to guess	7

Source: Survey data, 2005

5. Conclusions and Policy Implications

Compared to other socioeconomic challenges of Ethiopia, the household energy sector has attracted an insignificant share of concern and it has been off the development agenda. Besides, the development approach in Ethiopia has largely followed pro-rural approach. However, the implementation of this approach had been greatly hampered not only by the vulnerability of the environment, crop failure and declining land carrying capacity, but also by the emergent energy crisis. From the very modest fraction apportioned to urban sector, the lion share of development programs had been geared towards major urban centers than smaller ones. By and large, poor households in small towns have remained unattended.

Charcoal production impacts upon household economy (*through enhanced budget for energy*) and the environment (*through degradation*). Especially, the relationship between charcoal making and drought-induced poverty is demonstrated both for rural and urban settings. This contributes to a downward spiral of the poverty-environment nexus. This paper has provided an empirical contribution to the discussion on the linkage between charcoal production, poverty and environmental perception. The untenable nexus between poverty, declining soil productivity, environmental degeneration, and diminution of household energy found in the study area suggest immediate intervention in the form of fine-tuned policies and strategies.

This study has shown that a household energy in Ethiopia is highly dependent on biofuels, and the contribution of modern energy source is rather negligible. The heavy dependence on biofuels is hypothesized to have a detrimental impact on the environment, household economy and environmental awareness. Addressing the household energy problem of *Dugda Bora* has to be seen within the broader context of provision of sustainable livelihood for the various urban and rural communities that are dependent on biofuels.

The biofuel source base in *Dugda Bora wereda* has undergone massive degradation since the last 4-5 decades, not only from the demand generated by small towns in the locality, but also from ever-increasing demand generated from big cities like Addis Ababa. Currently the carrying capacity of agricultural lands has decreased and the vulnerability of the locality to drought has amplified through time. These phenomena further intensified urban poverty through the escalating cost of agricultural produce.

An important finding of the study is that charcoal availability in the urban centers is a function of seasonal food scarcity faced by rural households. As farmers are always living on the edge of poverty, even in good years, their agricultural produce cannot cover their annual food requirements. Hence during summer, when food scarcity reaches its zenith, the farmstead trees are used as sources of cash. Charcoal production would persist while knowing its impact on the overall environmental condition.

The study confirmed that charcoal making is a consequence of push and pull factors: poverty and drought on the other hand, and incentives from charcoal dealers on the other. Presently government policy seems to be insensitive to the dynamic link between environment and electric tariff. Under such circumstances whatever is endeavored to alleviate urban poverty, the success would be handicapped by the limitation posed by the scarcity of energy.

The current trend of energy use is unsustainable and there is an unattended "energy poverty" both in rural and urban areas. Its implication for sustainable development should be given increased prominence in the ongoing development debate. Given the strong nexus between degrading environment and increasing level of poverty in the study area, this paper recommends that the energy policy of the country has to be tuned to benefit the urban poor.

The widening demand-supply gap in household energy, coupled with the slow pace of making inroads into using modern energy sources, have made the problem of urban household energy to develop into a major challenge. Although low energy consumption is not a cause of poverty, the diminishing biofuel resource often correlates with various poverty indicators. Urban poverty is not an isolated phenomenon. Rather, it is intimately linked to environmental, economical, and social circumstances of the people concerned. However, such integratedness has rarely been given the limelight from researchers. Such dynamic relationship most often poses either a challenge or an opportunity in any attempt that is designed to benefit the urban sector. Solutions targeted at urban poverty should have ingredients from the rural context and this call for employing an integrated approach focusing on all facets of the urban problem.

An attempt should be made to create and diversify off-farm economic activity in the rural context. Another form of intervention is the delivery of microfinance services. The damaging demand for charcoal should be solved by a coordinated and strengthened control system encompassing big towns like Addis Ababa. The pro-poor policy agenda might include expanding access to grid electricity by amortizing upfront connection costs, sensitization of the public and politicians about energy issues.

Apart from the provision of amortization for installation of electric in urban areas, the existing electric tariff should be reconsidered based on a meaningful subsidy. In this regard, a cost benefit analysis should be made whether to export electric or promote rural electrification vis-à-vis environmental cost. Despite the ongoing argument that subsidies benefit the better off households through consumption of more electricity, one has to take into consideration of the insignificant proportion of better-off households in Ethiopia.

Moreover, alternative energy sources should be given special attention. Since woody biomass is declining and the electric rate is continuously rising, an alternative energy source becomes an urgent issue. In this regard, improved energy supply requires coordinated efforts across government sectors and administrative levels. Even in the rift valley, Wolde-Ghiorgis (2001) noted the existence of 700 MWe of geothermal energy potential, of which the Ethiopian Lakes district accounts for 170 MWe.

Given the importance of the energy sector in fighting poverty, there is emerging recognition by policy makers and NGOs that energy saving stoves (e.g., *Lakech*) can serve to address the challenge. In this regard, promoting and subsidizing the cost of energy saving stoves in urban and rural areas should be seriously considered. It is said that each *Lakech* stove saves an average of 75 kg of charcoal per household per year. This could enable savings comparable to tens of thousands of tones of charcoal. Such endeavor could save thousands of hectares of dry land forest in the region. However, due to poverty and a lack of appropriate alternatives, many will continue to rely on biomass for cooking. Hence the promotion of energy saving stoves should be accompanied with a massive afforestation program not only for the purpose of maintaining the environment, but also for the supply of firewood.

As a last resort, if food aid is necessary, there should be a well-timed system of delivery based on the rhythm of charcoal making in the rural areas. Besides, further studies should be undertaken in other charcoal producing areas to validate the result obtained in this study. The conclusions sketched in this study raise policy issues and the household energy problem needs continued debate across different sectors of the society.

References

Barnes, D. (1995). Consequences of Energy Policies for the Urban Poor. Energy Issues (FDP Energy Note No. 7).

Bereket Kebede., Almaz B. and Elias Kedir (2002). Can the Urban Poor Afford Modern Energy? The Case of Ethiopia. *Energy Policy* 30 (11-12): 1029-1045.

Bereket Kebede, Elias Kedir, Aselefech Abera and Solomon Tesfaye. (2004). Impact of Subsidies on the Affordability of the Modern Energy Services by the Urban Poor in Ethiopia: The Case of Electricity and Kerosene. African Energy Policy Research Network (AFREPREN) No. 322

Diamantini, C. and Patassini, D. (1996). Urban Ethiopia: Evidences of the 1980's.

Dierig, S. (1999). Urban Environmental Management in Addis Ababa: Problems, Policies, Perspectives, and the Role of NGO's. Hamburg African Studies, 8.

Douglas, B. (1995). Consequences of Energy Policies for the Urban Poor (accessed from internet)

Dasagupta, P. (1993). An Enquiry into Well-Being and Destitution, Clarendon Press.

EPA (1998). Background Information on Drought and Desertification in Ethiopia. Environmental Protection Authority, EPA, Addis Ababa.

Eckholm, E. (1984). Fuelwood: The Energy Crisis that Won't Go Away, In Eckholm, E., Foley, G. Barnard, G. Timberlake, L., Earthscan (Eds), London, England

Girma Kebede (2004). Living with Urban Environmental Health Risks: The Case of Ethiopia. King's SOAS Studies in Development Geography.

Hirut Bekele (2000). Natural Resource Degradation and the Predicament of Rural Women: The Case of Bugna *Wereda*, North Wollo. MA Thesis, Addis Ababa University, Regional and Local Development Studies (*unpublished*).

Hurni, H. (1988). Degradation and Conservation of the Resources in the Ethiopian Highlands. *Mountain Research and Development* 8(2/3).

Jogan, I. and Patassini, D. (1996). Urban Land Management in Ethiopia: the Case of Makalle.

Meikle S., and Bannister, A. (2003). Energy, Poverty and Sustainable Urban Livelihoods. African Energy Policy Research Network (AFREPREN) Working Paper No. 126.

Mesfin W/Mariam (1991). Suffering Under God's Environment: Vertical Predicament of Peasants in North Central Ethiopia.

Mesfin W/Mariam (1997). (*In Amharic*) The Choice: Poverty or Democracy? *Tobia Monthly Magazine*, 5 (7): 14-19.

MoFED (2002). Development and Poverty Profile of Ethiopia. Household Income Consumption and Expenditure and Welfare Monitoring Survey of 1999/00.

Pankhurst, R. (1992). The History of Deforestation and Afforestation in Ethiopia Prior to World War II. *Ethiopian Journal of Development Research*, 14 (2): 59-77.

Senait Regassa (1999). Household Energy Supply and Land Use in the Central Highlands of Ethiopia: The Choice Between Firewood and Cattle Dung. *Ethiopian Journal of Agricultural Economics*, 3(1): 89-119.

Woldeamlak Bewket (1995). Biofuel Consumption, Household Level Tree Planting and its Implication for Environmental Management in the Northwestern Highlands of Ethiopia. *EASSRR*, 21 (1): 19-38.

Wolde-Ghiorgis W/Mariam (2001). Renewables and Energy for Rural Development in Ethiopia: The Case for New Energy Policies and Institutional Reform. Short-Term Study Report, June 2001, AFREPREN Energy Research Program, Nairobi, Kenya.

Yacob Mulugeta (2003). Alternative Energy Technologies for Rural Development in Ethiopia. In: Assefa Abera, Getachew Tikub, and Johann B. (eds). Resource Management for Poverty Reduction: Approaches and Technologies. Ethio-Forum 2002.

2

Gender Dimensions of Urban Poverty in Ethiopia:
The Case of Three Kebeles in Addis Ababa

Meron Assefa

1. Introduction

1.1 General

Poverty is a multi-dimensional concept. Conventionally, it has been defined in terms of income or expenditure based on the assumption that a person's material standard of living largely determines their well-being. However, in recent years understandings of poverty and what constitutes well-being have been significantly broadened. It has been increasingly recognized that the conventional approach, though very imperative, fails to capture other significant aspects of individual's well-being such as the public provision of goods and services, access to common property resources and the intangible dimensions of life such as social relations, culture, security, and the natural environment.

On the other hand, the livelihood approach to poverty provides a broadened perspective as compared to the conventional approach in which the former has a holistic analytical framework providing a broad and systematic understanding of the various factors that restrict or enhance livelihood opportunities. "Livelihood refers to the ensemble of activities that a household (or an individual) regularly undertakes and the entitlements it makes claims to in order to sustain a given standard of living" (Dessalegn and Aklilu: 2002). Thus, the livelihood approach is concerned about not only the measurable income (consumption expenditure), but also about types of capital or assets upon which livelihoods are built and households or individuals strive to acquire in order to achieve positive livelihood outcomes. These assets consist of physical capital, the basic infrastructure and producer goods needed to support livelihoods; financial capital, the availability of cash or equivalent; human capital, the skills, knowledge, ability to labor and good health; natural capital, the natural resource stocks from which resource flows and services useful for livelihoods are derived; and social capital, networks and connectedness, institutions, and values.

Poverty or lack of access to these set of livelihood opportunities has its own distinctive features in space and time as well as among the various groups of the society within the same localities. In the urban context, poverty can be related to various distinctive characteristics of urban life and set of issues distinct from the general poverty situation. The World Bank has identified five dimensions of urban poverty: income/consumption, health, education, security, and empowerment which determine the livelihood of the poor (World Bank: 2002). Also the urban poor are a diverse group with varied needs as well as levels and types of vulnerability. These differences may be associated to factors such as gender, physical or mental disability, ethnic background, and household structure and also to the nature of the poverty itself (for example, long-term or temporary) (Ibid).

Thus, the extent of poverty appears differently among the various groups of society. It affects people of different characteristics in various ways for they have distinctive roles, needs and constraints. Hence, assessing urban livelihoods from a gender perspective would become crucial since men and women are often poor for separate reasons and have differing capacities to escape from poverty due to gendered constraints and opportunities in terms of access to income, resources and services.

To that end, the relationship between gender and poverty has been one of the issues that are debatable by practitioners in the area. The 'feminization of poverty', which argues that women bear unequal burden of poverty as compared to men, has been empirically explored. And in most cases many have asserted this notion to the extent that women have been singled out in efforts that address the problem of poverty. However, still a few practitioners doubt and question the appropriateness of the focus of poverty eradication efforts to address the gender dimension of poverty accrediting to the view that the causes and outcomes of poverty do not necessarily imply gendered context.

1.2 Problem Statement

Ethiopia is one of the poorest countries in the world. Using the various socio-economic indicators, one can see that quite a sizeable percentage of the population lives in abject poverty.

In the country; however, it is only recently that urban poverty has been recognized as an important area of concern. Studies focus on rural areas because the bulk of the population of the country lives in rural areas and also the rural population is frequently affected by recurrent famine and drought. Not only is the focus on rural poverty, but also literature shows that most of the studies emphasized primarily on the methodological measurements of poverty rather than examining the various aspects of urban poverty.

Urban livelihood in the country has become very complex in time confronting the urban people with various forms of deprivations. With the increasing rural-urban migration, the urban people have limited access to livelihood opportunities and hence income; inadequate and insecure housing and services; limited access to adequate health and education opportunities; and in general limited access to social and basic infrastructure.

On the other hand, it is obvious that men and women experience each of these challenges differently emanating from the reality that men and women play distinct roles, have varied needs, and encounter diverse constraints in various levels. Thus, the gender dimension of urban livelihood can have its own distinctive aspects as men and women experience and respond to urban poverty in different ways.

However, though much have been written about poverty in particular or gender issues in general, the relationships between the two have not been sufficiently assessed in the country. More importantly, the existing gender studies on poverty mainly deal with relating poverty and female headship. Due to reliance on household data, which are usually not gender disaggregated, these studies overlook the intra-household relations that are central to the analysis of poverty in gender perspective and focus on the outcomes rather than the causes of poverty.

This study is pertinent for various reasons. Firstly, it accentuates the multi-dimensional aspects of urban poverty and gender which are of pressing concern in the country. Secondly, also it attempts to bridge the research gap by adopting relevant methodologies. In order to overcome methodological limitations due to the reliance on household budget data, this study will employ a qualitative approach which will allow capturing the relevant variables that most explain the gender dimensions. And finally, an attempt will be made to provide fuller picture of the extent of the problem in order to contribute in terms of drawing consideration and prominence to the issue by providing more set of information.

1.3 Objectives of the Study

Overall, this study attempts to assess the gender dimensions of urban poverty in Ethiopia by taking the case of 3 Kebeles in Addis Ababa.

In particular, the major objectives of this study are to:
- identify the perception of the poor about poverty in the context of gender,
- assess the extent of urban poverty in terms of income, health, education, urban housing, and empowerment,
- identify the major factors contributing towards urban poverty from a gender perspective, and
- identify the coping strategies of the poor in the context of gender.

1.4 Methodology

The methodology adopted in this study is a qualitative one; it is based on a participatory approach using mainly focus group discussions and personal and key informant interviews. Such method is preferred to quantitative method of analysis for its appropriateness in exploring the multi-dimensionality of poverty from gender perspectives.

As it has been mentioned earlier, studies of poverty using the quantitative method and based on household data fail to take the intra-household distribution into consideration and are limited to mostly the female household head. Hence, the qualitative method is the most relevant, especially for our purpose since it enables us to explore broader concepts of poverty instead of focusing purely on the quantified household income/consumption levels and allow a better grasp of the multi-dimensional aspects of gender disadvantage.

In contrast to the conventional quantitative approach to poverty measurement, this method is helpful to measure and reflect people's perceptions and experiences. Such method enables one to look into intangible aspects of wellbeing or ill-being since it sets out to understand those intangible dimensions which do not lend themselves to measurement and hence are not detected in household surveys. Further, participatory methodologies also allow the direct involvement of people in problem identification.

The study examines a wide range of issues with a sample of urban residents in 3 communities in Addis Ababa: Kebele 19/20/21 (Guelele sub-city), Kebele 15/16 (Arada sub-city) and Kebele 11/12 (Kirokose sub-city).

The selection of these three sites is based on observations of localities which are commonly considered as being very poor. In selecting the study sites, the focus was to grasp the diversity of the poor communities, especially in terms of livelihoods. We were able to have discussions with group of men and women who are in petty trading, weaving activities, domestic house work, fuel wood carriers, commercial sex workers, elderly people, disabled people, students, jobless, and so on. Also to balance the information, key informant interviews were also held with officials from the Kebeles, health institutions and schools.

On the other hand, since the focus of this study is on the urban poor, poor people whose level of income was below Birr 1274 per annum (106 per month) were purposely selected for the study. This threshold is applied for it is the regional income/consumption poverty line for Addis Ababa applied by the government for its poverty assessment (MoFED: 2002b).

And finally, a set of questions in the form of a checklist was developed in order to facilitate the discussions.

2. Review of the Literature

Ethiopia being one of the poorest countries in the world, it is not arguable that the issue of poverty has to be high on the agenda of the country. Though it seems few in light of the severity of the predicament, there are assorted studies addressing the poverty situation of the country. Though different in terms of methodology and focus, the general conclusion of all the studies is the immense manifestation of poverty in the country measured by any standard.

In his assessment of the literature, Abbi (1997) revealed that poverty studies based on disaggregated household data in the country started to emerge as recent as 1995. Among other things, this is attributed to the availability of the household budget surveys: Welfare Monitoring Survey (WMS) and Household Income, Consumption and Expenditure Survey (HICES) since 1995/96 by the Central Statistical Authority (CSA); Ethiopian Rural Household Survey as well as the Ethiopia Urban Socio-economic Survey by the Economics Department of Addis Ababa University in collaboration with various external institutions which are available in 6 and 5 rounds respectively since 1994.

Despite various attempts, most of the poverty studies in the past focused mainly on developing the conceptual tools of measuring poverty (Mekonnen 1999a; Bigsten et al 1999a; Dercon and Mekonnen 1999; Abebe and Bereket 1996; Dercon and Krishman; Bevan and Bereket 1996; Bigsten and Negatu 1996). Many of the studies primarily address the controversy on the choice of welfare indicators, the derivation of the poverty lines and the choice of poverty measures (poverty indices). However, their significance is not deniable in terms of ascertaining the ground for poverty studies especially since the robustness of the study results depend, among other things, on the quality of the data and on how poverty is measured.

It is further noted that there are relatively few studies which deal with the determinants of poverty. According to Abbi (1997), in the early 1990's, no attempt was made (except a study by Mekonen in 1997) to analyze the factors that led to poverty. However, most of these studies are with primary emphasis on examining the econometric models that are most appropriate to model poverty in the country's context rather than providing thorough discussion of the poverty situation (Mekonnen 1999b; Abbi 1997; Bigsten et al 1999b).

Based on the survey data of 1995/1996 and 1999/00, the government has also produced two poverty profiles ('Poverty Situation in Ethiopia' and 'Poverty Profile of Ethiopia') in the year 1999 and 2002 respectively (MEDaC 1999; MoFED 2002b). These reports describe the poverty situation of the country, especially in urban and rural areas as well as in terms of the regional variations. Some of the results indicate that based on the data in the year 1995/96 about 45% of the population in Ethiopia is absolutely poor and unable to lead a life fulfilling the minimum livelihood standard. Further, significant difference in poverty levels between urban and rural areas has also been noted with about 47% of the rural population living in absolute poverty as opposed to only 33% of the urban population. (MEDaC: 1999). The subsequent poverty report, in the year 2002, further undertook inter-temporal analysis with that of the previous report. This description signified that using the 1999/00 household budget survey, the proportion of people who are absolutely poor (unable to meet their basic needs) was 44.2% in the country as whole, while it was 37% in urban areas and 45% in rural areas. The implication is that the number of people under the poverty line decreased by 3% at national level and by over 4% in rural areas while it increased by about 11% in urban areas (MoFED:2002b).

On the other hand, though the results of these reports are disaggregated spatially, both studies have no or very limited say on the gender dimensions of poverty. They further do not explore far beyond making comparisons on the poverty situation between male and female-headed households. The result suggests that in urban areas the number of female-headed households under absolute poverty is higher than male-headed households while there is no significant difference in poverty incidence (the number of people under the poverty line) between female-headed and male-headed households in rural areas. The reason for the later is argued that in Ethiopia most of the female-headed households in rural areas have land, which they can rent or manage themselves. On the other hand, it is further implied that in urban areas, though females can be engaged in income generating activities such as petty trade, they are usually involved in low paying activities. Besides, the fact that the literacy rate is lower for females than for males indicates that the capacity of females to generate income is still low (MoFED: 2002b).

Regarding changes in the poverty situation along gender lines, the report indicated that by 1999/00, poverty incidence has declined for male-headed households and increased for female-headed households compared to that of 1995/96. In terms of rural-urban perspectives, the extent of poverty has declined in rural areas for both male and female-headed households while in urban areas poverty incidence has increased for both. But the extent in the increase in poverty incidence is higher for female-headed than for male-headed households (MoFED: 2002b).

On the other hand, it is only a few studies that attempted to shed light on the gender dimension of poverty using the data from the household budget surveys (Girma 1997; Fistum 2002; Meron 2003). Most of these studies focus on comparing the extent of poverty among female and male-headed households and the correlates of poverty to female headship. The results suggest that female headship is not necessarily and directly correlated to poverty. Since these studies are restricted to analyzing the household survey data according to the differences in the gender of the household head, women can only be visible as female heads of households. Besides, the emphasis is on the outcomes rather than the causes of poverty.

However, intra-household relations themselves have been shown to be a powerful determinant of individual access to utilities and capabilities (Whitehead and Lockwood: 2000). The problem emanates since measuring poverty on the basis of household expenditure data effectively ignores the intra-household distribution as it is very rare to find standard surveys embarking on a quantitative exploration of intra-household poverty (Razavi: 2000). In applying household expenditure data, the easier ways to make gender visible is by dividing the households into male-headed and female-headed. This tendency to equate female headship with poverty has been refuted on both empirical and methodological ground (Razavi: 2000). It has been argued that disaggregating data by the gender of the household head does not provide a meaningful approach to gender and poverty (Whitehead and Lockwood: 2000). The reason is that the causes leading to female headship are clearly divergent, and the category of households labeled 'female-headed' is a highly heterogeneous one. It includes "lone female units, households of single women, wage earners with young dependents, households in which women earners receive significant remittances from absent males", and so on (Ibid). But aggregating these distinct categories of households generated through different social processes (e.g. migration, widowhood, divorce, etc) makes any simple comparison between male-headed and female-headed households impossible to interpret.

There are also studies which try to address the gender dimensions of poverty using qualitative data (Mulumbet: 2002). The study is more relevant in terms of better emphasizing and revealing the multi-dimensional nature of poverty from a gender perspective. The study investigates the peculiar aspects of poverty affecting women as compared to men and the reasons behind it.

Similar to the gender aspects of poverty, studies of poverty are limited in the urban areas. As Mekonnen (1999b) argued the focus has been on the rural areas for various reasons. One is that the bulk of the population lives in rural areas, and the intensity of poverty is immense looking at the recurrent starvations and drought in the country. However, because of the fact that in recent years poverty in urban areas is rising rapidly, the subject is attracting a growing interests (Mekonnen 1999b; Dercon and Mekonnen 1999; Mekonnen 1999a; Abbi 1997; Mekonnen 1996; Goitom 1996; Abdualhmid 1996; Bigsten et al. 1999; Meheret 2003; Shewaye 2002).

These studies explore the determinant of urban poverty with some of them making comparisons between urban and rural poverty. Abbi (1997), in modeling the determinants of poverty in Addis Ababa found that variables such as education, access to credit, employment status, gender, marital status and food shortage are significant determinants of welfare in Addis Ababa. Mekonnen (1999), in his exploration of the determinants and dynamics of poverty in urban Ethiopia, indicated that urban poverty is quite high with 46% of the population being under the poverty line in Ethiopia. There has been a high fluctuation in living standards of the people in urban Ethiopia mainly driven by fluctuations in food prices, particularly those of cereals. Bigsten et al. (1999) found that there are only small differences in urban and rural poverty levels. Rural poverty was significantly reduced between 1994 and 1997 while urban poverty remained largely unchanged. This has been also consistent with the results found by Dercon and Mekonnen (1999).

There are also some other attempts to look into other aspects of urban poverty. Meheret (2003) tried to explore urban poverty and urban governance institutions and pointed out that as Ethiopia is one of the fastest urbanizing nations, there is a need for the availability of strong urban governance institutions that can cater to the service and infrastructure needs of a growing urban community. Shewaye (2002) tried to shed light on the institutional capacities of urban Ethiopia to address urban poverty.

It is also high in the agenda of the development discourse of the country that the government of Ethiopia has designed a poverty reduction strategy paper (PRSP) entitled 'Sustainable Development and Poverty Reduction Program' (SDPRP). According to MoFED (2002a), the overall objective of this strategy is to reduce poverty through enhancing rapid economic growth while at the same time maintaining macroeconomic stability. Poverty head count ratio is projected to decline by about 10% by the end of the PRSP period (2004/2005) from its 1999/00 level of 44%. In light of this, it is stated that the government is committed to work towards meeting the Millennium Development Goals (MDGs) of 2015. And achieving such MDGs requires the Ethiopian economy to grow in real terms by 5.7% per annum until 2015 to reduce poverty by half from its current level.

The strategy paper also tried to shed light on the gender dimensions of poverty. According to MoFED (2002a), the government of Ethiopia is committed to eradicate poverty, particularly addressing its gender dimension. The document discusses that apart from endorsing the National Policy on Ethiopian Women, the government has created supportive constitutional provisions to establish women's equality with men. Realizing women's lack of access to extension services, low level of education attainment, and the need for gender sensitivity of the health system as well as greater vulnerability of women to HIV/AIDS, the government has claimed that it will be working towards addressing these issues (MoFED: 2002a).

In the governments annual progress report 2003/04 of the SDPRP, it is acknowledged that the government has moved decisively to advance the agenda on gender dimensions of poverty in the past year, and a significant number of initiatives are underway, including preparation of the National Action Plan on Gender as well as a gender budget analysis. A next step is to strengthen gender-disaggregated data reporting, in order to better inform policy.

To sum up, the aforementioned poverty literatures has a missing link when it comes to gender issues as many poverty studies have limited implication on the gender dimension. Besides, urban poverty has been of modest concern for many poverty studies in the country. Therefore, the overall review entails that much remains for further exploration in light of the enormity of the problem.

3. Major Findings of the Study

3.1 Community Profile

The study has been undertaken in 3 purposely selected Kebeles of Addis Ababa, Kebele 19/20/21, Kebele 15/16 and Kebele 11/12.

Kebele 19/20/21 is part of Gulele sub-city of Addis Ababa. It is located around Shero Meda, a northwestern suburb of the city. In an attempt to assess the variety of residents in the Kebele, the study examined the diversity of communities. Two types of communities stand out significant. There are those residing in what is commonly called 'Gamo Sefer'. This community is very isolated and has a rural setup. Gamo is the major ethnic group, Orthodox Christianity being the religion for most. All the men are engaged in weaving activities while the women are engaged in collecting and selling firewood. Most of the men are migrants from rural areas mainly Gamo Gofa zone of SNNPR seeking out better income by being engaged in weaving activities. Most of them migrate when they are young and after settling, got married at their place of origin and move to the city with their wives. Their language, Gamogna is widely spoken in the community. While many men speak Amharic, it is not the case for most women. The Gamo tradition and culture is still retained by the community. The other community studied in this particular site has an entirely different setup. The major ethnic group is Gurage, Tigre, Oromo, and Amhara while their major religion is Orthodox Christianity. Peculiar to this Kebele is that it is the northwestern boarder of the city, with Entoto forest nearby. Hence, though its significance is declining, collecting and selling firewood has been an important economic activity for many women.

The second site is in Kebele 15/16 of Arada sub-city. It is located around 'Arat Killo', which is nearly the center of the city. The major ethnic group in the area is Amhara and SNNPR while Orthodox Christianity is the religion of many people. Though the particular site is located at the center of the city, the community complained that it is isolated since public buses do not pass through the area. Also in the community, there are a lot of elderly and disabled people who were also the focus of the study. Due to its proximity to the center of the town, many commercial sex workers as well reside in the community usually renting a house in groups.

The last site studied, Kebele 11/12 in Kirokose sub-city, also located at the center of the city. The major ethnic group is Gurage, Oromo, and Amhara while Orthodox Christianity and to some extent Protestant and Islam are the major religion of the people. Exceptionally in this site, there are many female-headed households primarily due to death of their husbands.

3.2 Local Perceptions of Poverty

The participatory approach to poverty allows examining how poor people actually perceive poverty. This section explores a range of perceptions of the poor regarding many aspects of poverty.

The participants in various focus group discussions generated a number of definitions of poverty underlining its multidimensionality. The poor define poverty both in terms of material and non-material aspects that influence their livelihoods. Almost similar dimensions and aspects of poverty have been stressed across the three sites. Though poverty is ultimately defined in terms of lack of what is necessary for material well-being, it has been principally seen as the lack of livelihood resources or the lack of employment opportunities to generate income. Thus, poverty is perceived as lack of the means of livelihood and the inability to satisfy basic material and non-material needs. A range of related definitions of poverty, gathered from each focus group discussions, is given in Box 1.

In terms of material well-being, poverty is defined as inability to meet the financial needs to afford basic necessities, such as food and housing. Towards this money and purchasing power ability of the people are highlighted as significant determinants, attesting to the highly monetized urban economy. Inability to be engaged in income generating activities due to lack of working capital is also how some define poverty. Loss of livelihood due to illness and old age, and also the associated problems such as inability to afford medical care, or, in general, limited access to medical care are also defined as poverty. Poverty is also defined as illiteracy, and being uneducated in turn resulting in limited opportunities. Some also perceive poverty as not affording basic services such as housing, water and electricity in the community. Large family size is also described as impoverishment to some, especially the youngsters.

The non-material aspects of poverty is defined as a situation where one is living in a state of helplessness, anxiety and despair for being unable to work and meet the basic needs of the person and his/her family. It is defined as imprisonment and death where one is crippled to access the means to fulfill the basic needs.

Finally, what is implied in the discussions is that men and women reflect similar perceptions of poverty.

Box 2.1 Definitions of poverty by participants

"Lack of job opportunity"

"Lack of working capital and place to start income generating activities"

"Is like a prison where one can not fulfill his/her interest"

"Is like death or have no difference from death"

"Unable to use resources like labor"

"Being uneducated and illiteracy/or lack of access to education"

"Incapability (physical/mental inability) to work and generate income due to illness"

"Having no money/lack of finance"

"Unable to work due to old age"

"Lack of shelter"

"Having too many children"

"Not working"

"Laziness"

"Lack of market"

3.3 Income

In all of the sites, it was found that people are mainly engaged in informal, casual and daily wage labor, earning very meager income. Petty trading, selling food items ('injera' and bread), local drinks ('Tella' and 'Shameta'), vegetables, and charcoal; paid domestic works (mainly washing clothes and cooking); collecting and selling firewood and to some extent commercial sex work are the major sources of income for the women in the communities. On the other hand, the men are engaged in casual labor such as loading and offloading goods, cutting wood, dumping garbage, weaving activities and construction work. There are also few women engaged in these activities, which are traditionally considered as men's works. Thus, only a few people are engaged in permanent or salaried employment. These very few people engaged in low paying salaried jobs (mainly as guards, gardeners and pensioners), are considered to be better off in the community since they earn regular income.

The participants argue that they are limited to these low paying and insecure economic activities due to lack of jobs in regular formal employment and lack of other job opportunities in the city. They put the government as the culprit for failing to provide industries and other source of income, which could absorb the existing labor and improve their living conditions. Further, these problems were frequently attributed not only to lack of opportunities, but also to lack of capability due to limited knowledge and expertise to be competitive in the labor market. Participants are aware of their lack of knowledge of production and marketing, entrepreneurial and vocational skills (for instance, modern weaving methods for weavers) as limiting their income generating opportunities. On top of this, poor health and old age aggravate the already debilitating situations.

Limited livelihood opportunities endanger women into risky and socially demeaning activities such as commercial sex work and migrating overseas especially to Arab countries. Young girls in order to supplement their family's income and not to see the misery that their families are going through are motivated to get into commercial sex work (Box 2 gives a glimpse picture of their livelihood). Even the ones who are thought better-off in the communities manage to send their children aboard and supplement their income from the remittances.

Though the informal sector offers some opportunities for men and women to earn income, it has its own negative features that put them in a vulnerable situation. Lack of market place and financial resource are the major obstacles that the people who want to start petty trading face. Even the ones who are already in the business face such problems as expensive rental payment for market place ('Medebe') and inaccessible market places. Even though the Kebeles provide credit facilities through micro finance, the people seem to have limited access to these facilities. Limited awareness about the scheme and lack of collaterals were found to be the major reasons for men and women to be reluctant to acquire loans. The fear of the inability to pay back the loan and ending up being indebted is another stumbling block.

On the other hand, those women whose source of livelihood depends on fuelwood collection and selling encounter various problems, which basically emerge from the illegal nature of the work. Often women engaged in this work are exposed to the forest guard's harassment such as beatings and confiscations of the fuel woods collected as well as large amount of bribes for many forest guards in order to let them go with what they have collected. Since the place is unsafe, rape (especially of the young girls) by the forest guards and other men are the outstanding problems that the wood collectors face. Further, the heavy weight they carry and the long distance they cover coupled with the rugged topography of the hills have a daunting impact on the health condition of women. Despite all these difficulties, the fuelwood collectors finally face a market problem. They have very little bargaining power in the market as they come late from collections.

Also women whose livelihood depends on domestic work explained that they face maltreatments and harassments, which range from callous insults to rape at times. In addition to cutbacks and delays in payments, employers impose huge work loads to be completed in a very short time.

Box 2.2 The livelihood of commercial sex workers

The discussion with commercial sex workers revealed that among others, families inability to afford basic necessities and schooling, death of family member especially the head, lack of alternative opportunities, family disintegration, peer pressure and drug addiction as the major reasons forcing them into prostitution. One of the participants stated that she became a sex worker to escape from repeated rape attempts by her stepfather while some others said they are into this life at least not to be a burden on their families and some to support their families though almost all fail to do that. They remarked that it is better to earn one's bread than starving without doing anything.

Most of the time there are women fooled by brokers who lure them under the pretext that they will be waitresses but expose them to commercial sex work in bars.

Most of their earnings are spent on making themselves attractive with little left to cover their basic expenses of food, clothing, and rent for housing. One of the participants added that she failed to pay her rent in the past three months. They noted, "Business is slowing down". Their presumption is the men are becoming cautious about HIV/AIDS. Besides, there is an increase in the number of women getting into this life saturating the existing market.

The discussion revealed that they encounter a lot of problems threatening their lives. Those working at bars are mistreated and exploited by the owners of the bars. For instance, usually it is common to pay some money to the owner of the bar when the ladies go out with men. Those ladies working on streets especially described that they are exposed to physical abuses, violence, robbery, and drug abuse as well as car accidents and cut or denial of their payments.

The commercial sex workers felt that in most cases they are marginalized by the community. Since their way of life is condemned by their family or relatives, they usually lose communication. And the only time they might return home is when they are seriously sick. Some of them have migrated so far away from their homes so as not to be seen by friends, family or relatives.

As the nature of the work implies, they are seriously exposed to health hazards such as sexually transmitted diseases, HIV/AIDS and TB. Also threats to their health conditions are enormous. They noted that some men tear out the condom deliberately while having sex. They are also forced to perform sexual ill practices that could endanger their health. They are going through repeated abortion and at times have babies without any planning. The other illness commonly seen is what they traditionally call 'Boa' which is a dislocation of uterus. They remarked that it occurs from having too much sexual intercourse above their body's tolerance level. They usually get treatment for this particular illness from traditional healers since they do not get any cure in modern health institutions. And yet when they are sick, there is no one to care and support them since their friends are busy with their routines. The only option is to return to their families.

The men who are engaged in the casual labor also argued that they have limited job opportunities as the number of people who are engaged in such activities is growing rapidly.

The market problem is more pressing for the weavers. They remarked that it is the broker who has much better say for the product and the one who snatches the major profit. Further, they argued that the market price is very fluctuating and low as compared to the time and labor exerted to produce the item. As a result, many of them are utterly disappointed and moving away from the business and are sitting idle.

The source of livelihood for the old and the disabled is mainly the remittance (in kind and money) they get from relatives and the neighborhood (Box 3 summarizes the case study on the livelihood of elderly and disabled men and women).

> **Box 2.3 The livelihood of elderly and disabled men and women**
>
> Old and disabled men and women explained that due to their inability to work, they are left with no secured income and instead rely on the mercy of families/relatives and friends. These sources are very unstable and insecure which expose them to anxiety, despair and helplessness. Most of these women used to be engaged in petty trading and paid domestic work while the men were causal laborers.
>
> In addition, due to old age and their precarious living conditions, most of the elderly are subject to ill health that need much care and support. And yet these men and women lack people who can extend such care for them. Most of them pray for the sympathy of the neighborhood to get support. An old women described she will be taken out of her house once or twice a week with the help of her neighbors and she always wishes they are not tired of her and disappear.

In all the sites, it was found that women generate the bulk of the household income. This simply stems from the fact that they have better job opportunities than men in the labor market. In addition, out of desperation and in an attempt to fulfill their responsibilities, women are bound to step up to take low paying jobs than men do. One of the points raised by the participants was that the number of women in the informal labor force increases from time to time as compared to men though their income-earning opportunities are very weak. Further, women are engaged in casual labor, which were presumed to be men's work by the society in the past. Therefore, since most of the men are unemployed and underemployed in the community, households increasingly depend on women's incomes from jobs that are often considered undignified. However, this situation has its own consequences on the interaction of men and women in the household.

It is well known and confirmed by various studies that men's primary role is breadwinning and decision making as women's are to care for the family. This caveat has its deepest roots in the community and has to do with the tradition and norms of the society to the extent that it has become an identity that determines the gender division of labor. Consequently, due to their inability to contribute their share to the family income, the men feel that they are a total failure and a burden to the household. They are in desperation and agony as their roles have been taken away by the women in the community. However, through time, despite the rigid prescriptions of the traditional gender roles, the majorities of the men have taken the fact calmly, cope by cooperating and trying to internalize the fact for themselves. But it is worth also to mention that in some instances, such fact resulted in violence and family break-up or separations. This is especially true among the young couples.

It is widely seen that men leave the young girls after having babies. This is due to the insecurity and frustration that come out from their inability to support their families. Not only men, the women also are forced to leave their husbands due to the fact that their partners are not able to support the family. It is common to see young girls with one or two children moving back to their parents being separated from their partners for economic reasons. This might be the case even if women continue to care for their families; there is a threshold they are not able to go beyond as a sole breadwinner due to the insecure and low paying income-earning activities. Even some of the women hold their husbands responsible for not sharing the burden and could not stand living with them hence left them mostly with their children (Box 4 describes 2 such cases).

Participants described that the number of female-headed households in the community are large in number, especially due to death of partners, separation and abandonment. Most have a number of children and are faced with different challenges as compared to the married women.

Women's perception towards their role as the major breadwinner is mixed. Some believe that the women should remain at home and be responsible for household chores while her husband is in charge to generate income and bring it home. But some others, especially the young, believe that being active in the income generating activity has helped women to be exposed to the world outside her home (kitchen), to develop a sense of independence and also equally participate with the men in the household decision making. However, there is a sense of anxiety and cry of heavy burden is commonly shared amongst all. This is reflected by their shaky confidence on their capability to play such enormous roles in the household given their tenuous income-earning opportunities.

Box 2.4 Case studies of single women and female-headed households

Tigest Yemre, a female-head, aged 25 has a child with 15 months of age. She was separated from her husband as he was economically incapable to fulfill the family demands. He then went back to his family in rural area to be engaged in farming activity. But she refused to go with him and returned to her family's home with the newborn baby where she currently lives with her brothers. Since then, she has no contact with her husband. Her marriage was informal and there is no legal agreement signed.

She dropped out from school at grade 7 due to economic incapability as her mother died. She is now engaged in traditional female hairdressing to earn her living. She earns very small income as her work is very seasonal (better on holidays) and has no market in the locality. And yet she is managing the whole household and bears a lot of responsibly with regard to her child and also brothers.

She would like to find a permanent job but especially to expand the female hairdressing job by getting a working place and capital.

Kidist Kassa, 31 years old, with 2 children left her husband since he did not have a permanent job and could not fully support his family. He went to Gojjam to his family while she refused to follow him since her mother did not want her to leave. She then moved back to her mother but currently as her mother is dead, she is the head of the family.

She dropped out from school at 8th grade since she got a contract job but after the job was completed she failed to pursue her education. She is now engaged in petty trading mainly in vegetables and earns a small income of about 3 Birr per day. She is residing in a single room very suffocated with 7 family members sharing a latrine with 13 households.

The elder child (now 6 years of age) was attending a traditional church pre-schools ("Kese School") but now using the financial assistance she got from a neighbor, he is going to a kindergarten paying 40 Birr per month.

Her future dream is to find a job and able to send her children to decent school as well as complete her education at least up to grade 10.

In addition to their roles as the major breadwinner of the family, the study indicates that women still retain their primary responsibility for domestic chores. They are responsible for all the household work, health, education and well-being of their children and husbands. Women rise very early in the morning (on average around 5 a.m.), spend excessively long days and go to bed after all the household members. Ironically enough, despite all the work load women bear, they reflect their rigid belief in the traditional gender division of labor. They believe that household work is their natural duty that reveals their identity as a woman. Most of them strive to allocate their time and manage all the activities only by themselves since it is their natural responsibility. However, some argued that despite the firm belief in the traditional gender roles, few men help them with the domestic chores such as taking care of children, washing clothes, going to market, etc. Children (especially female) also have a major role in assisting their mothers with the household chores. Further, the women explained that they sometimes exhibit neglect in caring for their children, which emanates from excessive workload. A case study in the livelihood of women fuel wood carriers is summarized in Box 5.

Box 2.5 Women Fuelwood Carriers

The intra-household relations in the community where the fuelwood carrier women reside is quite different from the earlier noted. Most of the men are engaged in weaving as their main source of livelihood while the women are fuelwood carriers. The participants explained that even though the men are the major breadwinners, the women are also responsible to supplement the families' income by engaging themselves in the fuelwood collection and selling. Besides, they are responsible to all the households' chores in addition to assisting their husbands with the weaving activities that leave them with excessively long working days.

The women in general are not exposed to the outside world and are only confined to their day-to-day routine activities. Most of them do not speak other languages than their mother tongue since they do not go out of the house frequently. The women stated that giving birth to a female child is like a curse in their community back home. And that culture is still preserved even by those men who migrate to the cities. Women in general entertain the belief that women are created to serve men and rightfully remain to be dominated. The women had no chance to overcome such discriminations since they are still confined to their domestic activities and have no exposure to other experiences.

The participants were asked to prioritize activities from a list of activities they undertake. Most of them argued that one is not better than the other. Many are engaged in trading to overcome the day-to-day demands of life and not because it is a better deal.

In most of the sites, although the primary task of the children is believed to be going to school, in some instances it is found that children are engaged in various small income generating activities such as shoe shining, casual labor, street vending of plastic bags, taxi assistants ('Weyala'), etc. They are engaged in these activities after school and in some cases dropping out from school. Children are forced to supplement the household's income, especially to cover the cost of the children's education, clothing, and food. It is argued that such children perform low in their education and in majority of cases they end up dropping out of school. However, in one of the sites where the economic base is weaving, children are primarily engaged in the weaving activities and they do not go to school at all. This is especially true on those children who are brought from rural areas under the pretext that they will receive better education.

In most of the communities studied, women have strong social network to share burdens and workloads especially on childcare, exchanging labor during sickness, ceremonies, etc. However, unlike the usual expected tradition, having coffee together with close neighbors is not common among the women since their day time is taken up to meet their family needs with no time for other social activities.

Although the amount of income the households earned varies along the range of economic activities people are engaged in, on average it can be estimated to be 3 to 5 Birr per day. Further, such small income explains the food insecurity that the households face. The participants explained that since what they earn could not cover their food expenditures, they cope with the problem by reducing the quality, quantity, and frequency of their meals. Affording three meals per day is unattainable for most of the people and they would even be grateful if they get proper meal twice a day. One of the participants explained that most people in the community drink 'holy water' early in the morning. In addition to its religious value, people consider the 'holy water' as a breakfast and make them lose their appetites and not go hungry.

Their small earnings also limit them from affording other basic services such as education of their children, water, electricity, housing, etc. Especially, they argued that they have difficulties in covering their expenditures for 'Idir'. Most of the participants have an average of two "Idir's". It was found that being a member of 'Idirs' is immensely essential since they value the services they get mainly during the death of family members or themselves. For instance, an old woman with no stable source of livelihood and who earn her living by spinning cotton and selling it to weavers has 4 'Idir's' and pay 34 Birr per month while her earning is less than 100 birr per month.

Households in all the sites have large family size that is a result of many children and relatives who migrate from the countryside. This has considerable negative implication on the ability of the households to cover their living expenses. The participants explained that even if children and other members of the family are engaged in income earning activities, it is not significant enough to supplement the household's income. Thus, households face difficulties in raising their children and accommodating the migrants. Ironically, most of the households' attitudes towards large family size seem to be beyond their economical ability. Most of the households consider large numbers of children is one of the problems of their livelihood and yet they believe, on average at least, they should have 4 children. Even young couples who claim that they have difficulties in raising their children in the light of limited income and who use family planning, the number of children they think they can accommodate is large as compared to what they state they earn.

Urban life as the participants explained is getting more difficult as compared to past years. Increases in the cost of living resulting in decline in purchasing power of the households coupled with the decrease in livelihood opportunities has resulted in a huge gap in the livelihood of the poor. The situation is even thought to be worse for women since they are the ones who are responsible to bridge this gap and balance the livelihood of the households. The economic activities in the informal sector and causal labor works where majority is engaged in is narrowing down due to increases in the number of people entering the sector as it is the only easy way out for many.

Men and women argue that the government has to build industries to improve the problem of limited income earning opportunities. Such industries should absorb the inactive labor and should accommodate people of various levels of education and age. Besides, they propose that government should help them increase their capabilities by providing them with various skill trainings (marketing, entrepreneurial and vocational). Women participants argued that government should avail credit facilities on individual rather than on group. Further, provision of accessible market places is identified as one of the priorities.

41

3.4 Empowerment

The discussions revealed that, while it seems men have the upper hand in household decision-making, this was not always the case. One major issue in this regard is related to the decision on income and spending. Here, where the research was conducted, women were found to be breadwinners of the household. They are mainly involved in petty trading activities and earn some income and thus have the authority to manage their money. It is the women who make daily household decisions and monitor major expenditures like purchase of clothes, food, etc. However, due to their little income, both men and women do not have that much room to maneuver.

In most cases, women understate and even fail to notify the amount they earn to their husbands. They are compelled to such behavior just to avoid inquiries of money by husbands for personal use such as drinking, chewing chat and the like. In the few instances where husbands are engaged in income generating activities, it is highly likely that they spend a significant portion of it for their personal consumption (like gambling, smoking and the like). What came out clearly from the discussions is the fact that women give more priority to family well-being as compared to men. However, other decisions concerning children's education, health and family planning are made jointly.

In all communities (except the one where fuelwood carriers reside) women and men participants stated that there are no abuses directed against women such as rape, domestic violence, women trafficking and abduction. They also added that harmful traditional practices such as early marriage, female genital mutilation and the like are virtually nonexistent. The participants attributed this to the increased role of women as a breadwinner hence economic independence, and increased awareness.

However, some of the participants argued that there are a few cases of domestic violence, which they define as beating and insults. But still, both men and women perceived domestic violence, especially beating wives, as acceptable. One old woman argued that it is customary for a husband to beat his wife since men have the tendency to lose their temper very easily. This clearly shows domestic violence is a deeply rooted cultural practice. But the youngsters in the discussion groups seem to believe that such attitudes are fading away. They noted that if a man beats his wife, they will report him to the Kebele and the police. The key informant interview with Kebele officials revealed that there are some cases of domestic violence and rape reported to them at the Women's Affairs Desk. Most of the cases are resolved at an early stage by the Kebele officials and community elders while a few cases have been brought before the Social Affairs Court in the Kebeles.

Gender based abuses especially rape, domestic violence, abduction and early marriage are highly manifested among the community of Gamo in Kebele 19/20/21. Most of the women got married at a very early age of 15 and migrated to the city with their husbands. Rape is very common, the victims being those who are engaged in wood collection. The women showed reluctance to report the case since they believe it would not bring about long-term solution. Besides, they feel that in their quest for justice their children will be made to endure suffering and abandonment.

As part of the discussions on empowerment, participants were asked about their attitude towards the mass media (newspaper, radio, TV, etc.). The people have very little attachment to media sources and this is even more pronounced among women. Women participants explained that since they are burdened by excessively long working days, they have no time for the media. Even if the radio is on, the women feel that they have very little concentration to listen to the programs since they are possessed by many thoughts and are mentally and physically exhausted. They even prefer the radio not to be on since they feel disturbed by the noise given the tedious day they spend. However, the men are relatively keen to follow the news and other programs. The youngsters are usually tuned to the FM station.

Participants were asked how easy it was for them to express their views in front of local officials and what kind of responses and the reactions they get from them. With some exceptions, both men and women participants noted that they have adequate knowledge of their rights as citizens and the standard of service that the Kebele and other authorities render. They can freely express their views to the local officials; in this regard, things have improved lately, in the past 6 months. However, they are often dismayed since none of their demands have been fulfilled by the concerned bodies. Due to this, many of them have stopped to go to the Kebeles. For instance, they have repeatedly requested improvement in housing and sanitation, but so far the officials have turned a deaf ear to their requests. Residents in 'Gamo Sefer' especially feel that they are marginalized and completely forgotten by the Kebele officials.

Participants were also asked if they are familiar with various women-focused programs and the duties of the Women's Affair Desk as well as the organized women's association by this desk in Kebeles. Most of the women are familiar with various programs working for women such as microfinance; awareness creation on health related issues such as family planning and HIV/AIDS (door to door service); resolving disputes between husband and wives; etc. However, the participants in the discussion groups as well as interviews with the Kebele officials revealed that the participation of women in the women's association is very minimal even if large numbers of women are registered as members. This is attributed to women's inability to pay monthly contribution (4 birr per month), their negative attitude towards the institution and lack of adequate time to be active. Even some of those participating in the association feel it has no benefit at all.

Further, participants were asked how the formal and informal laws serve men and women in the community. They explained that in the formal law, men and women are equally treated but in the informal law (traditional/cultural) women's rights are not well honored. A common case is that elderly people always side to the men when resolving family disputes. Yet, most of the people in the community use such laws at least at early stages of dispute.

3.5 Health

In all of the sites studied, government health institutions, especially health centers and clinics, are the most commonly used. Fortunately, these institutions are located in close convenience to the communities in question. However, participants in "Gamo Sefer" complained that their health center is a good distance from their locality, bad thus one has to pay for transportation to receive treatment. Besides, some centers are ill-suited for transportation and not easily accessible to patients.

Despite the availability of contiguous health institutions, the institutions have problems such as insufficient medicines, high treatment costs, nepotism and discrimination of health staffs, and limited capacity, which has limited the majority of the population from seeking treatment there.

Almost all the participants could not afford the medication costs which ·include consultation fees (only 1 Birr in government health centers), treatment charges, drug costs and others. Hence, many obtain free medical care upon producing the necessary certification from their Kebeles. The participants stated that nowadays it is very easy to get certifications from their Kebeles.

However, all the participants argued that they are usually mistreated because they are non-paying patients. The government pharmacies in the health centers are usually out of drugs leaving them with no options but to return home since the private pharmacies are exorbitant. The participants strongly doubt the claim that government pharmacies are short of supplies and question the staffs' personal integrity.

The participants also recognize that the quality of treatment is low in some instances. They attribute this to the large number of patients and feeble motivation of the health staff. Also they believe that the city health bureau has loose monitoring and supervision of its employees. Further, they characterize the service at times as discriminatory and callous. Some participants argue that it emanates from overload of duties and exhaustion. But most allege that the preferential treatment goes to those who pay and the better off. They also mentioned that the health officials are constantly in urgency, which dwindle proper consultation leading to prescribing the same drugs for every case before laboratory examination.

The health officials also asserted that though the health centers have the standard number of health workers allotted by the Ministry of Health, the number of patients visiting each day is beyond the capacity of the institutions. This overload is manifested in terms of the long queues of people seeking health service and the enforced quota limiting the number of patients to be treated per day. The later is especially the case for adults above the age of 14 while no quota is imposed on children under 14. In one of the sites (Kebele 11/12), a patient must start to queue at 4 a.m. in the morning to get the service. Even if the capacity of the health center is to treat as large as 130 to 140 adult people per day, it is apparently customary to see large number of patients denied service as their number exceeds the allotted quota.

As the overall situation is a major deterrent to access health institutions, it is not surprising to see the majority of the population turning away from formal institutions. Most of the inhabitants visit the centers in quest for the free drugs. Some even usually take medicines without consultations of doctors or medical officers. On top of that, many consider these institutions as a last resort and visit when their infirmity is terminal. Rather, 'holy water' is the first line treatment to the majority for its affordability and easy access coupled with its quintessential religious value. However, the community only makes use of traditional healers when they have skin infection such as 'Almaz Bale'chera'.

Private health facilities are able to ease the pressure on government health institutions on a small scale. Though many perceive that these institutions render better services, their high treatment cost have restricted many from visiting them. And those who make use of private health institutions usually rely on remittances from relatives and neighbors and even sell assets to meet the costs for the service.

It is confirmed by the participants and key informants from health institutions that in terms of physical facilities, the government health centers and clinics are well-equipped and staffed up to the standard of the Ministry of Health. Rather, lack of enough space for medical service and office work is the major problem cited. There are situations where medical treatment is given publicly, infringing the privacy of the patient.

Health officials and residents of the communities agreed that most of the common diseases affecting the community are caused by the poor environmental sanitation which stems from lack of latrines, poor drainage system, improper waste disposal and overcrowded living area. The most common illnesses mentioned include pneumonia, cough, cold, TB, asthma, diarrhea, typhoid, skin infections and dysentery, which are ascribed to the poor living condition as well as malnutrition. Children are frequently affected by cough, cold and diarrhea.

In addition to this, most women are exposed to diseases related to unsafe working conditions and work-related risks, such as burns (domestic workers), skin infections (especially weavers), backache, bone fractures and fall injuries. The participants argued that most fail to get medical care since they consider they would not get a lasting solution as long as they stay in the same occupation. Further, associated to their living conditions, women argued that they are exposed to mental stress, anxiety, and depression as they worry too much about their routine works.

In all the sites, participants perceive that the number of people especially women affected by HIV/AIDS is quite large. But due to their fear of stigma and discrimination most sufferers do not divulge their disease. Especially, if a person loses weight, or is affected by TB or "Almaz Bale'chera", the community surmises the individual as an AIDS patient. On the other hand, many of the participants took HIV/AIDS test while some others are afraid to take the test as they envisage the discouraging aftermath if they are found to be positive. Participants argue that people are aware of Voluntary Counseling and Testing (VCT) services at the government health centers. And they increase their awareness about the disease using the disseminated information on posters, door to door awareness creation programs, mass media, etc.

Participants argue that poor health status of a member of a family immensely affects the livelihood of a household. In all the sites, people disabled due to illness cannot be engaged in any income generating activities hence are a burden to their families, relatives and neighbors.

The government health centers give free family planning services. Women participants explain that they get the service from the institution adequately with much care and support. Awareness creation on family planning is given door to door by Kebeles and concerned NGOs. The communities' attitude towards family planning is positively altering. And the numbers of the beneficiaries from such schemes are intensifying across time. Paradoxically, there is quite a large number of young girls with children out of wedlock.

Antenatal, delivery and postnatal care services are provided by the government health centers, but delivery will be referred to hospitals if there is any pregnancy complication. While the number of women getting antenatal and postnatal care is quite high, most women fail to get delivery service in government health centers. This is largely due to the fact that as this service is just at an initial stage in these institutions, women prefer giving delivery at government hospitals and some at their homes. Besides, some of the participants wrongly presume they would not get delivery service in the institutions if they had no prior antenatal visits in the same institution. Due to this, the health centers at the sites studied are operating under capacity. In addition, it is well noted that some women just prefer to give birth with the assistance of non-medically trained women at home. The predicament is prominent if they confirm that they have no pregnancy complication during their antenatal visit.

3.6 Education

The community commonly uses government schools located in the surrounding area. In most of the sites studied, there are nearby government elementary and high schools. However, access to education is constrained among others, by the socio-economic factors affecting the community. Although both men and women perceived education is a key for the well being of individuals, parents are in some cases reluctant and in most cases unable to send their children to school. As a result, most children are denied access to school and absenteeism as well as dropouts is high (Box 6 describes a case study of school dropouts).

Even if students in the primary (Grade 1 to 8) and high school (Grade 9-10) pay no tuition fees, they are expected to contribute some sum of money for development of schools while for preparatory schools (Grade 11 to 12) students pay registration fee. Most children were said to be out of school because parents could not afford the school fees and contributions.

46

The parents noted that they are required to contribute money for development purposes, which do not take their ability to pay into account. The amount of money for the contribution (mostly ranges from Birr 75 to 150) is usually proposed by the committee which combines parents, teachers and the school administration. This is determined in a meeting with all the parents present. However, in all the sites, parents complain that the contributions always do not mull over the ability of families to pay and the decisions are made in haste. Besides, they complained they are required to contribute more than once in a year, which makes it difficult for them to plan and program the payment. One parent said that, "When we feel relieved paying this one, the other one comes in no time."

The school officials, on the other hand, argue that this rate is determined by considering the poor economical status of the community and the complaints have little to do with ability to pay. They remarked that most households give little value to their children's education. Besides, in some cases in order to ease the burden, contributions are made only for one child even if the family has more in schools. But this privilege is provided if the children are attending schools administered by the same Kebele. However, this claim was refuted by the participants in the group discussions. One of the participants said that all her three children are now out of school since she can not afford the contribution imposed by the school (100 birr per child). Her elder son is now engaged in daily labor to support the family.

Box 2.6 Case study of school dropouts

W/o Etalem Mengistu is a single mother of 4 children (2 female and 2 male). She is very old and her youngest child is 20 years old. None of her 4 children completed their education due to the family's inability to support them.

While he was alive, her husband was a priest with no permanent job. He used to earn a small income by teaching small children and also later managing public water stand in the vicinity while she was engaged in collecting and selling firewood along with some petty trading. Yet all this was not enough to support the family and keep on sending the children to school. The family suffers from shortage of food and poor shelter which is too old to protect them from rainfall and flood. The situation was terrible putting so much pressure on the children that all except the youngest daughter abandoned the family feeling that the street would not be worse than their home. All of them discontinued their education. The boys started living on the streets working as causal labor while the girl tried to leave the country illegally.

Her husband finally died 9 years ago. She finally managed to get a permanent job as a laborer at the City Municipality and earns 200 Birr per month. Later, she tried to bring back her family once again and managed to get them all back home. Her house was maintained with the support of relatives, the Kebele and neighbors. The 2 boys joined the army during the Ethio-Eritrea war while the girl got married and left the family. Even if she tried hard to send her youngest daughter to school, her daughter dropped out of school (at 6th grade when she was 16 years old) since she was abducted and got pregnant by one of the friends of her brothers.

Currently, W/o Etalem Mengistu is supporting 6 people: her two sons who are retired from the military with one of them with a wife; her youngest daughter with her baby as well as one of her relatives who came from the rural area. Two of her sons continue their education at the night shift, but her youngest daughter could not manage it since she has nobody to take care of her baby while she is at school. Her two boys also support the family to some extent through causal labor.

Most of the participants also stated that they could not afford the cost of school materials and uniforms. Many parents believe uniforms are essential to discourage clothing competition among students. However, a few parents claim that it is an additional school expenditure since they would have sent their children to school with their casual clothes. School officials assert that nowadays most students manage to come to school wearing uniforms though most of the students wear worn-out and unclean uniforms due to the inability and largely disregard of parents to their children.

Further, since schools are well distributed and are within walking distance, transportation cost is not a barrier to education in most of the cases. However, there are many cases where students are assigned to preparatory schools (11th to 12th), which are very far from the students' homes. For instance, a student living at Shiro Meda area was assigned to Nifas Silk to attend the 11th grade which is very far from his home and cannot cover the transportation cost; he had no other option but to discontinue his education.

Another significant barrier to sending children to school is the demand for child labor to assist the household in house work and in income generating activities. This is more evident for females and those who are brought from rural areas. Boys tend to drop out of school to work as taxi assistants ('weyala'), shoe shining, and causal labor and in some cases because they are keen on sports. There are a very few cases where they run away from school to engage in drugs and violence. Female students, especially those who come from the rural areas, absent themselves from class to undertake household chores. Also some female students drop out or are absent to help with income generating activities. One child of 7th grade stated that if her mother has some social engagements, it is her responsibility to undertake the petty trading on that particular day and be absent from class.

The other major setback for the unsatisfactory enrollment in schools is the reluctance of parents to send their children to schools. This is owing to the increasing unemployment for the urban youth and the dissatisfaction arising from it. Consequently, most are in fear of their children's future. Also for those who are currently enrolled in school, this has caused deep frustration and agony. In one discussion, a student said that her parents always asked her "What difference would you make if you get education or not?"

The frequency of girl students attacked by hooligans has been reduced as compared to the past. Parents and teacher associations are functioning in all schools though they are very inactive since members are busy with their personal affaires and lack incentives. The main purpose of the associations is to assist in the school administration, create close relationship between the school and the community, and take discipline measures on both teachers and students.

3.7 Housing

The majority of the participants reside in rented Kebele houses. However, a few people, especially those in the peripheral areas, live in privately owned houses. There are also some squatter settlements mainly at the peripheral areas. The participants argued that

48

though rents for Kebele houses are relatively cheaper (on average Birr 5 to 10 per month), many fail to pay their rents regularly. This issue is very pressing for those who live in private rented houses. They complained that the rent is far beyond what they can afford. Though many people have repeatedly applied for Kebele houses, they did not get positive responses.

Many of the participants are in perpetual fear of eviction by the government as part of the demolishing process underway in certain areas of the city. The participants expressed their insecurity since this would worsen their precarious livelihood situation. This is especially because moving from their neighborhood would result in losing their income generating activities since most of their activities are carried out around their homes and the customers are people from the neighborhood. Above all, it can also damage their strong social networks such as 'Idir' and 'Iqubs', they have put together for many years. Others also complained that they cannot afford the newly government built houses (condominiums) that they would be offered in exchange. Even though many people know that they cannot afford the condominiums, they have applied and registered for them in the Kebeles (Box 7 summarizes one case of displaced people currently residing in a temporary shelter).

Most of the houses are small sized, with single or two rooms which are used as a living room, bed room, kitchen and the like. And yet many people (more than 8 in some cases) live in each of these overcrowded and suffocating dwellings. In some places, especially among the weavers, the rooms are also used as a working place. Most of the houses are dilapidated and could not properly protect the residents from rain. Further, the residential area is very overcrowded with dwellings sharing the same wall and roof. Such condition exposes the residents to various communicable diseases. The participants also argued that the crammed houses are a fire hazard. In addition, sounds can easily be heard between the houses invading the privacy of the residents.

The sanitation problems are also very severe. Some people use shared (communal) latrines though the numbers of households who share the toilet are beyond the capacity of the latrines. Some of the latrines are in poor condition with weak floors and collapsed walls and are also very close to the houses affecting the people nearby with foul smells. There are also a large number of people who use public toilets that is available in or nearby the community. A good fraction of the people have no access to any type of sanitation outlets thus are forced to use river sides, fields/forests and even containers like buckets and plastic bags and dispose them afterwards. For instance, the residents in 'Gamo Sefer' have no access to any type of toilet at all; hence, the people use forests, rivers and their backyard as their means of sanitation.

Box 2.7 Case study of displaced people

The residents of the community have been removed from their original home since the area was reputed by the government to be a hiding place of criminals. The houses were illegally constructed adjacent to a small market area though later the Kebele recognize the people as its residents and issued them residents' identification cards. The houses were demolished with the promise to provide other houses (condominiums) in exchange within six months; the promise has not been fulfilled and the six months have already passed putting many into frustration. Some of the residents of this community are still living in plastic shelters at the old place since they were not given temporary shelters.

The temporary houses are constructed by the displaced residents using the remains of the construction materials from their demolished houses on land provided by the Kebele. One of the participants stated that there is a continuous fight over a sleeping place in her family. In her former place, she used to have 4 rooms for a family of eight. Some of her children still sleep in the old place in a plastic shelter. Besides, most of the houses fail to properly protect from rain and flooding. The people use the road sides as places for cooking. In order to use the nearby community latrine, they were obliged to contribute 70 Birr each. Since many could not afford this, they are forced to use the public toilets which are very far and unhygienic.

Many of these people used to carry out their economic activities, mainly petty trading near their old homes and their buyers were their neighbors. Now all have lost their source of income and are worried as to how they are going to support their families. During the discussion, some participants were in tears about their situation which said was no better than death. Some still hope that they might get condominiums though they doubt its affordability.

Solid waste is disposed mostly in the nearby hillside and rivers. Participants complained that there are no waste disposal containers or burying and burning wastes is also not possible in the vicinity. Hence, they are exposed to bad smelling garbage piled up at the doors of their homes. People are also exposed to the wastes dumped in the nearby rivers. Further, rain brings serious environmental hazard as it pushes away all the wastes from the rivers to the streets exposing the community to serious health dangers. Due to the lack of proper sewage system, most of the sites (especially Kebele 11/12) are subject to frequent floods during rainy seasons. Participants in this Kebele stated that the rainwater floods into their houses and causes physical damages and loss of lives. This particular site is very prone to flooding and there were instances where people were carried away by floods.

Participants in all the sites complained that they do not have access to individual pipe water connections; hence they must either purchase water from privately owned taps or public stands ('Bono'). The cost (money, labor and time) is considerably higher than what they would incur if they had private connections. This has implications on the people's livelihood since it weakens businesses in the informal sector such as food processing and making and selling local drinks that depend on reliable supply of potable water. Besides, it adds more burdens on women as they are responsible for household chores and also are mainly engaged in the informal sector. Women complained that even though they are in urban city, they have to collect water from public stand pipes located at a distance from their homes, spending some time due to the long queue. Girls are especially responsible for collecting water, which may affect their education negatively. Further, households also do not have direct electric connection instead they share light sources with others. Due to its high cost, people do not use electricity for other household purposes. The major source of cooking fuel is firewood and kerosene.

4. Conclusion and Policy Implications

4.1 Conclusion

This report presents results from a qualitative study carried out in Kebeles in Addis Ababa. The overall objective of the study was to assess the extent of various dimensions of urban poverty with a gender perspective. The research employed participatory approach that basically involved focus group discussions, as well as personal and key informant interviews. Basically, five variables, income, education, health, empowerment and housing, which explicate the urban dimensions of poverty, have been the core of the discussions.

The poor defined poverty as multidimensional which has both material and non-material aspects. Both men and women perceive poverty as lack of livelihood opportunities that deprive one's ability to fulfill basic needs. Further, the findings revealed that people are faced with multi-faceted problems of livelihood. However, due to our reliance on qualitative data, it would be difficult to make a definitive conclusion about the relationship between gender and poverty. The information gathered showed that in some instances women are worse off than men while in others they appear to be better off.

However, we may argue that women are highly burdened by poverty due to their greater role in household's income earning activities. It was found that women are the major breadwinner of the household. This is the case since women have better access to jobs and also they are desperate to fulfill the family's demand by stumbling upon any activities, even traditionally considered as men's work. This adds to the burden of women as they still are responsible for domestic chores, but also has a positive impact in improving their decision making power in the household. Paradoxically, despite the greater burden, women still believe in the traditional gender division of labor. Further, the repercussion of this shift of traditional gender roles sometimes causes family break-ups and separations, especially among young couples.

The study showed, for both men and women, that lack of secured and permanent livelihood is a key concern as the majority relies on low-paying and insecure work in the informal sector. Women are mainly engaged in petty trading, paid domestic work, and collecting and selling firewood while the men are engaged in unreliable causal labor. Even these works are diminishing due to the influx of people entering the informal sector as it is easy way out for many. Due to this, some women are forced to do risky and socially demeaning activities such as* commercial sex work, migrating overseas and (illegal) fuel wood collection. Besides, the majority of men and women stated that their precarious livelihood conditions are getting worse due to persistent increase in cost of living and decreases in purchasing power as well as declining livelihood opportunities. Large family size depresses the ability of families on the ability to cover the living expenses.

Children are also engaged in income earning activities to supplement the household's income. The findings disclosed that a significant number of children are denied access to education while absenteeism and dropouts are also common, especially among female students who come from rural areas. The major reasons are the inability to cover the costs of education, demand for child labor to assist in household work and income generating activities as well as parents' reluctance to send children to school as many remain jobless after completing their high school education.

The discussion groups indicated that domestic violence and other forms of abuse on women are decreasing which is attributed to the improved economic role of women as breadwinner, and increased awareness. Although many women are familiar with the various women-focused programs of the Kebele, their participation in the activities is insignificant. Many women as compared to men have very little attachment to media outlets since they are burdened by excessively long working days.

Since Kebele (government) rented houses are the major form of affordable housing, men and women are in fear of eviction from their homes by the government because of the demolishing process underway in some areas of the city. While the rent for Kebele houses is perceived as affordable, there is more concern for those who reside in private rented housing because it deprives them of their scanty earnings. The majority lives in small sized single room houses, which are dilapidated and do not properly protect from the rain. Over crowdedness of dwellings, shortage of latrines and waste disposal outlets as well as poor drainage systems are the main causes for the poor health of the majority.

Men and women commonly use government health institutions, usually producing certification for free treatment. However, many complained about the shortage of drugs, the limited capacity of health institutions, and the discrimination and nepotism by health staffs.

4.2 Policy Implications

Various multi-faceted and interlinked factors explain the manifestation of poverty on both men and women. These multiple dimensions of deprivation necessitate a broad perspective to engender change in the livelihood of the urban poor.

In view of the findings, various pertinent policy issues may be raised for discussion. However, my focus here is on how to make the most of the existing resources to the best benefit of the community rather than to suggest ambitious plans for change. With this context, the subsequent interventions are of high priority.

Enhancing income generating opportunities: Men and women stressed the lack of income generating activities as the main obstruction to their livelihood. Thus, in the long term, the prime emphasis should be to expand the bases of the economic sectors in order to be able to generate alternative income earning activities that can absorb the existing inactive and insecure labor force. Most importantly, in the short-term, support should be extended to improve the informal sector which absorbs the majority of the poor. As the poor themselves put it, facilitating the various microfinance credit schemes and mitigating the problems associated with market places, entrepreneurial, and other vocational skills would be important areas of focus. Addressing these problems would especially ease the burden of the large number of women in the sector that are distressed by being the sole bearer of all the responsibilities in the household.

Adequate interventions to improve housing conditions: The study has shown the people in the three Kebeles live in poor housing conditions; have pressing problems of environmental sanitation; and face personal insecurities associated with the evictions from their homes. Moreover, the livelihood of the poor is closely interlinked with their place of living. In view of this, government interventions should avoid threatening the poor with loss of their scanty livelihood and the importance of their social networks. Also much effort should be exerted toward improving the poor environmental sanitation that is an immense health hazard to the majority of the people.

Emphasis on the education of children: The findings emphasized children's access to education is jeopardized by the precarious living condition of the poor. In particular, the unpopular money contributions for school development should not be imposed on the poor. Rising awareness of men and women of the value of education despite the cost they incur would also improve the insight of the poor to let their children go to school. However, the long term solution would be improving the livelihood of the poor to reduce the role of children in supplementing household income.

Strengthen family planning programmes: Though the people's perception on the benefits of family planning has improved, still most have large families, which worsen their already deprived living condition. Therefore, a much more strengthened family planning programme that can eliminate the already inherent belief in large numbers of children should be the focus of concerned bodies.

Need to properly utilize the government health institutions: Men and women complained about access to the government health institutions. However, this situation should cease to be an impediment rather concerned bodies should ensure the utilization of the institutions to the best benefit of the community. To that end, proper service delivery mechanisms and monitoring instruments should be employed to guarantee the standard service the institutions are supposed to render.

References

Abbi Mamo, (1997), "Modeling Poverty and its Determinants in Addis Ababa: A Focus on Multinomial Logit Selection Model", *Ethiopian Journal of Economics*, Vol. 2.

Abdualhmid Bedri, (1996), "Poverty and Nutritional Status in Urban Ethiopia", in: Bereket Kebede and Mekonen Tadesse (eds), *Poverty and Economic Reform in Ethiopia*, Proceedings of the Fifth Annual Conference on the Ethiopian Economy.

Abebe Shemeles and Berket Kebede, (1996), "Issues in the Measurement and Dynamics of Poverty: A Survey", in: Bereket Kebede and Mekonen Tadesse (eds), *Poverty and Economic Reform in Ethiopia*, Proceedings of the Fifth Annual Conference on the Ethiopian Economy.

Bevan, Philippa and Bereket Kebede, (1996), "Measuring Wealth and Poverty in Rural Ethiopia: A Data-based Discussion", in: Bereket Kebede and Mekonen Tadesse (eds), *Poverty and Economic Reform in Ethiopia*, Proceedings of the Fifth Annual Conference on the Ethiopian Economy.

Bigsten, Arne and Negatu Mekonnen, (1996), "Some Results on the Level and Distribution of Income in Urban Ethiopia", in: Bereket Kebebde and Mekonen Tadesse (eds), *Poverty and Economic Reform in Ethiopia*, Proceedings of the Fifth Annual Conference on the Ethiopian Economy.

Bigsten, Arne et al., (1999a), "Poverty and Welfare in Ethiopia: Profile and Determinant", *Ethiopian Journal of Economics*, Vol. VIII, No. 1.

Bigsten, Arne et al., (1999b), "Changes in Welfare and Poverty: an Application of Stochastic Dominance Criteria", *Ethiopian Journal of Economics*, Vol. VIII, No. 1.

Dercon, Stefan and Krishnan, Pramila (1996), "A Consumption-based Measure of Poverty in Rural Ethiopia in 1989 and 1994", in: Bereket Kebede and Mekonen Tadesse (eds), *Poverty and Economic Reform in Ethiopia*, Proceedings of the Fifth Annual Conference on the Ethiopian Economy.

Dercon, Stefan and Mekonene Tadesse, (1999), "A Comparison of Poverty in Rural and Urban Ethiopia", *Ethiopian Journal of Economics*, Vol. VIII, No. 1.

Dessalegn Rahmato and Aklilu Kidanu, (2002), "Livelihood Insecurity among Urban Households in Ethiopia", Discussion Paper No. 8, Forum for Social Studies.

Fistum Tehaye, (2002), "Poverty in Addis Ababa: A Comparison of Female and Male-headed Households", M. Sc Thesis, Addis Ababa University, (Unpublished).

Girma Seifu., (1997), "Poverty in Addis Ababa: A Comparison of Female and Male-headed Households", M. Sc Thesis, Addis Ababa University, (Unpublished).

Goitom Ghirmatsion, (1996), "Aspects of Poverty in the City of Addis Ababa: Profile and Policy Implication", in: Bereket Kebede and Mekonen Tadesse (eds), *Poverty and Economic Reform in Ethiopia*, Proceedings of the Fifth Annual Conference on the Ethiopian Economy.

Mekonnen Tadesse, (1996), "Food Consumption and Poverty in Urban Ethiopia", in: Bereket Kebede and Mekonen Tadesse (eds), *Poverty and Economic Reform in Ethiopia*, Proceedings of the Fifth Annual Conference on the Ethiopian Economy.

Mekonene Tadesse, (1999a), "Perceptions of Welfare and Poverty: Analysis of Qualitative Responses of a Panel of Urban Households in Ethiopia", *Ethiopian Journal of Economics*, Vol. VIII, No. 1.

Mekonene Tadesse, (1999b), "Determinants and Dynamics of Urban Poverty in Ethiopia", *Ethiopian Journal of Economics*, Vol. VIII, No. 1

Meron Assefa, (2003), "Female-headed Households and Poverty in Urban Ethiopia", M. Sc Thesis, Addis Ababa University, (Unpublished).

Ministy of Economic Development and Cooperation (MEDaC), (1999), "Poverty Situation in Ethiopia", Addis Ababa.

Ministy of Finance and Economic Development (MoFED), (2002a), "Ethiopia: Sustainable Development and Poverty Reduction Program", Addis Ababa.

-------------------- (2002b), "Development and Poverty Profile of Ethiopia", Addis Ababa.

Mulumebet Zenebe, (2002), "An Analysis of Household Poverty from a Gender Perspective: a Study Based on two Kebeles in Addis Ababa", in Meheret Ayenew (ed.), *Poverty and Poverty Policy in Ethiopia*, Forum for Social Studies, Consultation Papers on Poverty No. 7.

Razavi, Shahra, (2000), "Gendered Poverty and Well-being: Introduction", in Razavi, Shahra (ed.), *Gendered Poverty and Well-being*, UNRISD Discussion Paper No. 94.

Shewaye Tesfaye, (2002), "A Review of Institutional Capacity to Address Urban Poverty in Ethiopia", In Meheret Ayenew (ed.), *Poverty and Poverty Policy in Ethiopia*, Forum for Social Studies, Consultation Papers on Poverty No. 7.

Whitehead, Ann, and Lockwood, Matthew, (2000), "Gendering Poverty: a Review of Six World Bank African Poverty Assessment", In Razavi, Shahra (ed.), *Gendered Poverty and Well-being*, UNRISD Discussion Paper No. 94.

World Bank, (2002), "Urban Poverty", In Baharouglu D., and Kessides C., (eds), *A Source Book for Poverty Reduction Strategies*, Vol. 2.

FSS Publications List

FSS Periodical

Medrek, now renamed BULLETIN (Quarterly since 1998. English and Amharic)

FSS Discussion Papers

No. 1 *Water Resource Development in Ethiopia: Issues of Sustainability and Participation.* Dessalegn Rahmato. June 1999

No. 2 *The City of Addis Ababa: Policy Options for the Governance and Management of a City with Multiple Identity.* Meheret Ayenew. December 1999

No. 3 *Listening to the Poor: A Study Based on Selected Rural and Urban Sites in Ethiopia.* Aklilu Kidanu and Dessalegn Rahmato. May 2000

No. 4 *Small-Scale Irrigation and Household Food Security. A Case Study from Central Ethiopia.* Fuad Adem. February 2001

No. 5 *Land Redistribution and Female-Headed Households.* By Yigremew Adal. November 2001

No. 6 *Environmental Impact of Development Policies in Peripheral Areas: The Case of Metekel, Northwest Ethiopia.* Wolde-Selassie Abbute. Forthcoming, 2001

No. 7 *The Environmental Impact of Small-scale Irrigation: A Case Study.* Fuad Adem. Forthcoming, 2001

No. 8 *Livelihood Insecurity Among Urban Households in Ethiopia.* Dessalegn Rahmato and Aklilu Kidanu. October 2002

No. 9 *Rural Poverty in Ethiopia: Household Case Studies from North Shewa.* Yared Amare. December 2002

No.10 *Rural Lands in Ethiopia: Issues, Evidences and Policy Response.* Tesfaye Teklu. May 2003

No.11 *Resettlement in Ethiopia: The Tragedy of Population Relocation in the 1980s.* Dessalegn Rahmato. June 2003

No.12 *Searching for Tenure Security? The Land System and New Policy Initiatives in Ethiopia.* Dessalegn Rahmato. August 2004.

FSS Monograph Series

No. 1 *Survey of the Private Press in Ethiopia: 1991-1999.* Shimelis Bonsa. 2000

No. 2 *Environmental Change and State Policy in Ethiopia: Lessons from Past Experience.* Dessalegn Rahmato. 2001

No. 3 *Democratic Assistance to Post-Conflict Ethiopia: Impact and Limitations.* Dessalegn Rahmato and Meheret Ayenew. 2004

Special Monograph Series

1. *Lord, Zega and Peasant: A Study of Property and Agrarian Relations in Rural Eastern Gojjam.* Habtamu Mengistie. 2004.

FSS Conference Proceedings

2. *Issues in Rural Development. Proceedings of the Inaugural Workshop of the Forum for Social Studies, 18 September 1998.* Edited by Zenebework Taddesse. 2000

3. *Development and Public Access to Information in Ethiopia.* Edited by Zenebework Tadesse. 2000

4. *Environment and Development in Ethiopia.* Edited by Zenebework Tadesse. 2001

5. *Food Security and Sustainable Livelihoods in Ethiopia.* Edited by Yared Amare. 2001

6. *Natural Resource Management in Ethiopia.* Edited by Alula Pankhurst. 2001

7. *Poverty and Poverty Policy in Ethiopia.* Special issue containing the papers of FSS' final conference on poverty held on 8 March 2002

Consultation Papers on Poverty

No. 1 *The Social Dimensions of Poverty.* Papers by Minas Hiruy, Abebe Kebede, and Zenebework

Tadesse. Edited by Meheret Ayenew. June 2001

No. 2 *NGOs and Poverty Reduction.* Papers by Fassil W. Mariam, Abowork Haile, Berhanu Geleto, and Jemal Ahmed. Edited by Meheret Ayenew. July 2001

No. 3 *Civil Society Groups and Poverty Reduction.* Papers by Abonesh H. Mariam, Zena Berhanu, and Zewdie Shitie. Edited by Meheret Ayenew. August 2001

No. 4 *Listening to the Poor.* Oral Presentation by Gizachew Haile, Senait Zenawi, Sisay Gessesse and Martha Tadesse. In Amharic. Edited by Meheret Ayenew. November 2001

No. 5 *The Private Sector and Poverty Reduction [Amharic].* Papers by Teshome Kebede, Mullu Solomon and Hailemeskel Abebe. Edited by Meheret Ayenew, November 2001

No. 6 *Government, Donors and Poverty Reduction.* Papers by H.E. Ato Mekonnen Manyazewal, William James Smith and Jeroen Verheul. Edited by Meheret Ayenew, February 2002.

No. 7 *Poverty and Poverty Policy in Ethiopia.* Edited by Meheret Ayenew, 2002

Books

1. *Ethiopia: The Challenge of Democracy from Below.* Edited by Bahru Zewde and Siegfried Pausewang. Nordic African Institute, Uppsala and the Forum for Social Studies, Addis Ababa. 2002

Special Publications

• *Thematic Briefings on Natural Resource Management. Enlarged Edition.* Edited by Alula Pankhurst. Produced jointly by the Forum for Social Studies and the University of Sussex. January 2001

New Series

• Gender Policy Dialogue Series

No. 1 *Gender and Economic Policy.* Edited by Zenebework Tadesse. March 2003

No. 2 *Gender and Poverty (Amharic).* Edited by Zenebework Tadesse. March 2003

No. 3 *Gender and Social Development in Ethiopia.* (Forthcoming).

No. 4 *Gender Policy Dialogue in Oromiya Region.* Edited by Eshetu Bekele. September 2003

No. 6 *Gender Policy Dialogue in Southern Region.* Edited by Eshetu Bekele. December 2004.

• Consultation Papers on Environment

No. 1 *Environment and Environmental Change in Ethiopia.* Edited by Gedion Asfaw. Consultation Papers on Environment. March 2003

No. 2 *Environment, Poverty and Gender.* Edited by Gedion Asfaw. Consultation Papers on Environment. May 2003

No. 3 *Environmental Conflict.* Edited by Gedion Asfaw. Consultation Papers on Environment. July 2003

No. 4 *Economic Development and its Environmental Impact.* Edited by Gedion Asfaw. Consultation Papers on Environment. August 2003

No. 5 *Government and Environmental Policy.* Consultation Papers on Environment. January 2004

No. 6 የግልና የጋራ ጥረት ለአካባቢ ሕይወት መሻሻል (የሰሜን ሸዋ ግራሮች ተምክሮ) Consultation Papers on Environment. May 2004

No. 7 *Promotion of Indigenous Trees and Biodiversity Conservation.* Consultation Papers on Environment. June 2004

• FSS Studies on Poverty

No. 1 *Some Aspects of Poverty in Ethiopia: Three Selected Papers.* Papers by Dessalegn Rahmato, Meheret Ayenew and Aklilu Kidanu. Edited by Dessalegn Rahmato. March 2003.

No. 2 *Faces of Poverty: Life in Gäta, Wälo.* By Harald Aspen. June 2003.

No. 3 *Destitution in Rural Ethiopia.* By Yared Amare. August 2003

No. 4 ***Environment, Poverty and Conflict.*** Tesfaye Teklu and Tesfaye Tafesse. October 2004

Educational Materials

- Environmental Posters
- Environmental Video Films
    - የከተማችን አካባቢ ከየት ወደየት?
    - ውጥንቅጥ፣ ድህነትና የአካባቢያችን መጎሳቆል

Africa Review of Books (Managed by FSS for CODESRIA)

- Vol. 1 No. 1. October 2004.

ከድህነት ወደ ልማት፦ ዕውቀትን ለትውልድ ማስተላለፍ

- ቁጥር 1፡ ድርቅንና ረሀብን ለመቋቋም የተደረጉ እንቅስቃሴዎች (1966-1983)። ሽመልስ አዱኛ፣ 1997 ዓ.ም.

www.ingramcontent.com/pod-product-compliance
Lightning Source LLC
Chambersburg PA
CBHW080844270326
41929CB00016B/2914